Advances in Comparative and Environmental Physiology 15

Volumes already published

Vol. 1: NaCl Transport in Epithelia. Edited by *R. Greger* (1988)

Vol. 2: From the Contents: Hibernation – Vapor Absorption. Nutrient Transport – Stable Pollutants (1988)

Vol. 3: Molecular and Cellular Basis of Social Behavior in Vertebrates. Edited by *J. Balthazart* (1989)

Vol. 4: Animal Adaptation to Cold. Edited by *L. C. H. Wang* (1989)

Vol. 5: From the Contents: Osmoregulation – Chemoreception – Temperature – Spectroscopy – Metalloproteins – Prostaglandins (Eicosanoids) – Maternal-Fetal Relationships (1989)

Vol. 6: Vertebrate Gas Exchange: From Environment to Cell. Edited by *R. G. Boutilier* (1990)

Vol. 7: From the Contents: Protein Turnover – Xenobiotic Metabolism Lipoproteins (1991)

Vol. 8: From the Contents: Dormancy in Aquatic Invertebrates-Root Effect – Red Blood Cell Functions in Fish-Diving Mammals (1991)

Vol. 9: Volume and Osmolality Control in Animal Cells. Edited by *R. Gilles, E. K. Hoffmann,* and *L. Bolis* (1991)

Vol. 10: Comparative Aspects of Mechanoreceptor Systems. Edited by *F. Ito* (1992)

Vol. 11: Mechanics of Animal Locomotion. Edited by *R. McN. Alexander* (1992)

Vol. 12: Muscle Contraction and Cell Motility: Molecular and Cellular Aspects. Edited by *H. Sugi* (1992)

Vol. 13: Blood and Tissue Oxygen Carriers. Edited by *Ch. P. Mangum* (1993)

Vol. 14: Interaction of Cell Volume and Cell Function. Edited by *F. Lang* and *D. Häussinger* (1993)

Vol. 15: From the Contents: Salivary Gland Secretion – Nematocyst Discharge – Adaptations in Decapodan Crustaceans – Steroid-Sensitive Areas Mediating Reproductive Behaviors (1993)

Volumes in preparation

Biomechanics of Feeding in Vertebrates. Edited by *V. L. Bels, M. Chardon,* and *P. Vandewalle*

Ion Transport in Vertebrate Colon. Edited by *W. Clauss*

Electrogenic Chloride Transporters in Biological Membranes. Edited by *G. A. Gerencser*

Effects of High Pressure on Biological Systems. Edited by *A. G. Macdonald*

Mechanisms of Systematic Regulation. Vol. I: Respiration and Circulation. Vol. II: Acid-Base Regulation, Ion Transfer and Metabolism. Edited by *N. Heisler*

Advances in

Comparative and Environmental Physiology 15

With Contributions by

G. F. Ball · J. Balthazart · M. Hidaka · I. Novak · M. Wheatly

With 24 Figures

Springer-Verlag
Berlin Heidelberg GmbH

ISBN 978-3-642-77530-7 ISBN 978-3-642-77528-4 (eBook)
DOI 10.1007/978-3-642-77528-4

ISSN 0938-2673

Softcover reprint of the hardcover 1st edition 1993

Production Editor: Herta Böning, Heidelberg
Typesetting: Best-set, Hong Kong, and K+V Fotosatz GmbH, Beerfelden
31/3145-5 4 3 2 1 0 – Printed on acid-free paper

Foreword to the Series

The aim of the series is to provide comprehensive, integrated reviews giving sound, critical, and provocative summaries of our present knowledge in environmental and comparative physiology, from the molecular to the organismic level.

Living organisms have evolved a widespread range of basic solutions to cope with the different problems, both organismal and environmental, with which they are faced. A clear understanding of these solutions is of course of fundamental interest for all biologists, zoologically or medically oriented. They can be best comprehended in the framework of the environmental and/or comparative approaches. These approaches demand either wide surveys of animal forms or a knowledge of the specific adaptive features of the species considered. This diversity of requirements, both at the conceptual and technological level, together with the fact that physiology and biochemistry have long been mainly devoted to the service of medicine, can account for the fact that these approaches emerged only slowly amongst the other new, more rapidly growing disciplines of the biological sciences.

The field has now gained the international status it deserves and the organization of a series devoted to it appeared timely to me in view of its actual rapid development and of the interest it arouses for a growing number of biologists, physiologists, and biochemists, independently of their basic, major orientation.

Liège, Belgium, Spring 1988 *Raymond Gilles*

List of Editors

Editor-in-Chief

R. Gilles
Laboratory of Animal Physiology
University of Liège
22, Quai Van Beneden
4020 Liège, Belgium

Editorial Board

P. J. Butler
School of Biological Sciences
University of Birmingham
P. O. Box 363
Birmingham B15 2TT, England

R. Greger
Physiologisches Institut
der Universität Freiburg
Lehrstuhl II
Hermann-Herder-Strasse 7
7800 Freiburg, FRG

Ch. P. Mangum
Department of Biology
The College of William & Mary
Williamsburg, VA 23185, USA

G. N. Somero
Department of Zoology
Oregon State University
Corvallis, OR 97331-2914
USA

K. Takahashi
Zoological Institute
Faculty of Sciences
University of Tokyo
Hongo, Tokyo 113, Japan

R. E. Weber
Zoophysiology Laboratory
University of Aarhus
Building 131
8000 Aarhus C, Denmark

Contents

Chapter 3

Physiological Adaptations in Decapodan Crustaceans for Life in Fresh Water
M. Wheatly

Chapter 4

Neurochemical Differences in Two Steroid-Sensitive Areas Mediating Reproductive Behaviors
J. Balthazart and G.F. Ball

Chapter 1

Cellular Mechanisms of Salivary Gland Secretion

Ivana Novak

Contents

1 Introduction

The vertebrate salivary glands, unlike most other exocrine glands, show a bewildering variety of structural and physiological patterns (Young and van Lennep 1978, 1979). This variability is exhibited not only among the various types of salivary glands (e.g., parotid, mandibular and sublingual), but also in homologous glands among different species. It is thought to reflect the diversity of dietary habits and adaptations of different species. Salivary glands secrete digestive enzymes such as amylases, proteases and lipases. The role of enzymes in normal digestion may be limited, but these would gain in significance in cases of pancreatic insufficiency such as in cystic fibrosis (Valdez and Fox 1991). Generally, a more important function of saliva may be to moisten and clean the buccal mucosa and to act as a lubricant to aid swallowing, and in some species it is used for evaporative

Physiologisches Institut, Albert-Ludwigs-Universität, Hermann-Herder-Str. 7, W-7800 Freiburg, FRG

Advances in Comparative and Environmental Physiology, Vol. 15

cooling and grooming (Young and van Lennep 1978; Valdez and Fox 1991). The salivary glands also produce various hormones, homeostatic and growth factors such as glucagon, erythropoetin, renin, kallikrein, nerve and epidermal factors and other polypeptides (Young and Schneyer 1981) (see Sect. 1.2). These biologically active substances may be involved in regulation other than those associated with digestion, but proof for such functions is still outstanding.

After a brief note on the historical perspectives of salivary gland physiology and a description of some basic morphological features of the major glands, the main emphasis of this chapter will be to review the events which lead to the secretion of saliva, especially the membrane mechanisms involved in electrolyte and water transport. The common transport mechanisms in secretion, absorption and their regulation will be described, and where known, usually from studies on intact glands, differences between various glands and species will be pointed out.

1.1 Historical Perspectives

The study of salivary gland physiology could be said to have begun when Thomas Wharton (1659) and Niels Stensen (1662) identified and described the duct of the human submandibular and parotid glands, respectively. Prior to this discovery, it was not appreciated that salivary glands could form an external secretion. Rather, the glands were treated as mysterious organs which behaved somewhat like sponges straining off "excrementous substances" from the blood, such as bad spirits of the brain. For almost another two centuries it was believed that blood vessels communicated somehow with the interior of the glands, and all sorts of ingenious explanations were given to account for the different types of secretions formed by various glands (see Garrett 1975; Young and van Lennep 1978 for reviews of early history). This belief was held until Ludwig (1861) and Brettel (1869) showed that the salivary glands did not belong to the same functional group as the kidney, since they could form a secretion in response to nervous stimulation. Moreover, the secretion was formed independently of the arterial blood pressure. In the following century, a theory was proposed that the secretion is formed by the active transport of solute, which by building up an osmotic gradient results in the passive movement of water. Heidenhain (1878) was interested in the mechanisms involved in the secretory process. He found that the secretion of organic matter was exhaustible with time, but the secretion of salts was not. Studying the relation between the rate of secretion and the composition of saliva, he observed that the salt content of saliva varied in proportion to the secretory rate, i.e., stimulation of the salivary glands usually results in the production of a hypotonic secretory product. The concentration of electrolytes in saliva and, therefore, the total osmolality vary with the secretion rate, so that saliva is usually hypotonic when collected at low secretory rates and nearly isotonic when collected at high rates of secretion. This phenomenon remained unexplained for more than half a century. Thaysen and colleagues (Thaysen et al. 1954; Thaysen 1960) studied this relation between the salivary secretory rate and the salivary composition of major electrolytes. As a result the two-stage hypothesis of salivary formation was enunciated (Thaysen 1960). In the following years this hypothesis was tested rigorously by attempting to analyse the function of endpieces and ducts separately in micropuncture and microperfusion experiments (see Sects. 2.2 and 3). The essential points of the extended and modified hypothesis which apply to most, but not all, glands are: (i) the secretory endpieces produce an isotonic primary saliva, which is plasma-like in composition; (ii) this primary fluid, as it passes along the salivary duct system towards the exterior, is modified by the processes of Na^+ and Cl^- reabsorption in excess of K^+ and HCO_3^- secretion, such that the final saliva becomes hypotonic. This hypothesis is basically descriptive and does not explain how secretory endpieces and ducts perform their transport. This issue of cellular transport mechanisms will be taken up in the following sections.

1.2 Morphology

The comparative morphology of salivary glands has been well described in a monograph and a recent review (Young and van Lennep 1978; Young et al. 1987). Hence, in this section I shall briefly outline some basic features of salivary gland morphology, in order to provide a framework for the rest of the review. Salivary glands can be described as being composed of at least two functionally and morphologically separate types of epithelium (Young and van Lennep 1978, 1979). They are compound glands, consisting of secretory endpieces draining into a series of branched ducts. The terminal parts of the parenchyma, or so-called endpieces, are thought to produce and discharge the main fluid and organic components of the secretion. Endpieces contain one or more of the following cell types: serous, mucous or seromucous, named according to the proportion of protein/carbohydrates in the secretory granules. These cells are arranged in acinar or tubular structures or combinations of these. In most mammals the secretory endpieces of parotid glands appear histologically and histochemically similar, while often the mandibular gland endpieces may consist of more than one cell type. For example, the rabbit mandibular gland has secretory endpieces containing central serous tubules with terminal and peripheral seromucous acini. In contrast, the rat mandibular gland has a simpler endpiece consisting only of seromucous acini (Young and van Lennep 1978). Often, secretory endpieces are simply referred to as acini irrespective of their actual shape. Ultrastructurally, the main features of secretory cells are well developed rough endoplasmic reticulum at the basal pole of the cytoplasm and numerous secretory granules at the apical pole.

An obvious function of the series of convergent ducts is to convey the formed saliva, but they can also modify the secreted fluid by reabsorbing and secreting various electrolytes and by adding some organic solutes to the fluid. Secretory endpieces drain into intercalated ducts. From earlier ultrastructural studies it was assumed that there is a relatively simple relationship between the secretory endpieces and intercalated ducts. Recent studies, e.g., on human and dog parotid glands, show that one intercalated duct may serve many acini, some of which drain directly into it, while other neighbouring acini first join an extra short segment of intercalated duct (Dardick et al. 1990). Interestingly, in some glands (e.g., hamster mandibular), first- and second-order intercalated ducts contain secretory granules (Chaundry et al. 1985). Intercalated ducts join the endpieces to the striated ducts, the striated appearance of which is due to the alignment of mitochondria in the cytoplasmic processes formed by extensive infolding of the basal and basolateral membrane. Greatly modified striated ducts with a large number of secretory granules are found in the mandibular glands of some rodents, such as the rat and mouse. According to their appearance, these ducts have been named the granular ducts. Here, adult animals exhibit sexual dimorphism. In response to testosterone and thyroid hormone, males develop an extensive granular convoluted ductal system, the cells of which synthesize unusually high amounts of epidermal growth factor (Watson et al. 1985; Royce and Baum 1991). These cells also contain high amounts of nerve growth factor and renin, which are released as a result of aggressive behaviour (Bing and Poulsen 1979; Watson et al. 1985; Aloe et al. 1986) and in response to α-adrenergic stimulation (Hazen-Martin and Simson 1987). The striated and granular ducts drain into the intralolubar and extralobular (interlobular) excretory ducts, which in monostomatic glands eventually join a main excretory duct. The salivary endpieces and, to a lesser extent, the intercalated ducts are surrounded by basket-like, spindle-shaped cells resembling smooth muscle cells (Young and van Lennep 1977, 1978; Garrett and Emmelin 1979). It is thought that these so-called myoepithelial cells prevent distension of the endpieces during secretion and provide propulsive force to aid the expulsion of secreted fluid.

Generally, all three major salivary glands receive a rich supply of nerves from both divisions of the autonomic nervous system, parasympathetic and sympathetic, which supply the endpieces, ducts, myoepithelial cells and blood vessels (Young and van Lennep 1978; Garrett 1982). In the late 1970s, in the wake of immunocytochemistry, it became apparent that a variety of peptides, such as vasoactive intestinal peptide (VIP) and substance P are found in the autonomic nerve endings (Hökfelt et al. 1977; Uddman et al. 1980; Lundberg 1981; Olsen et al. 1986). What their physiological roles in salivary glands are is unclear, but it has been shown that substance P, related tachykinins and VIP are capable of eliciting secretion in some glands (Coroneo et al. 1979; Martinez and Martinez 1981; Inoue and Kanno 1982; Lundberg et al. 1982; Ekström 1987; Wright et al. 1991). Various purines have also been considered as neurotransmitters (Gallacher 1982; Petersen 1987; Soltoff et al. 1990).

2 Secretory Endpieces

2.1 Volume of Secretion

It is presumed that the bulk of saliva is formed in the secretory endpieces (see Sect. 1.1), but so far no satisfactory measurements of these rates have been possible. Hence, the function of the salivary endpieces has been estimated from measurements of end saliva volumes. In most salivary glands no secretion is apparent unless the glands are stimulated. However, there are glands in some species which exhibit a so-called spontaneous secretion, defined as secretion occurring in the absence of any apparent nervous or humoral stimuli (Babkin 1944). The best known example is the sheep parotid and glands of other Bovidae (Coats et al. 1956; Kay 1960), where such secretion can reach up to 15% of the maximum stimulated rate. Other glands including the cat and dog sublingual glands (Emmelin 1953) and rabbit mandibular gland (Young et al. 1981; Novak 1984) secrete spontaneously at a lower rate, e.g., in the rabbit this amounts to about 1% of the stimulated secretion. Actually, it is most likely that many salivary glands form primary spontaneous secretion, and therefore the collection of primary saliva by micropuncture has been possible. However, since small volumes may be reabsorbed as saliva passes down the reabsorptive duct system, no secretion might be detected at the usual site of collection – the main duct.

Copious saliva is usually secreted in response to stimulation of the autonomic nerves to the glands or by agents mimicking the action of autonomic neurotransmitters. All salivary glands are innervated by parasympathetic secretomotor fibres, and many but not all also by sympathetic secretomotor fibres (Young and Schneyer 1981). Parasympathetic stimulation produces vigorous fluid secretion, which is usually greater than that produced by sympathetic stimulation (Fig. 1). Sympathetic stimulation, however, evokes secretion that is richer in organic components (Young and van Lennep 1979). The high rates of secretion evoked by sustained maximal stimulation are usually maintained for brief periods, and then secretion decreases to a plateau, as seen for example in the rat and rabbit mandibular glands (Fig. 1; Murakami et al. 1986; Novak and Young 1986). Such a biphasic response depends on the dose of the agonist, or the mode with which the agonist sets the stimulus-secretion reaction in action, e.g., choline causes little tachyphylaxis in the rat mandibular gland in comparison with acetylcholine (Murakami et al. 1986).

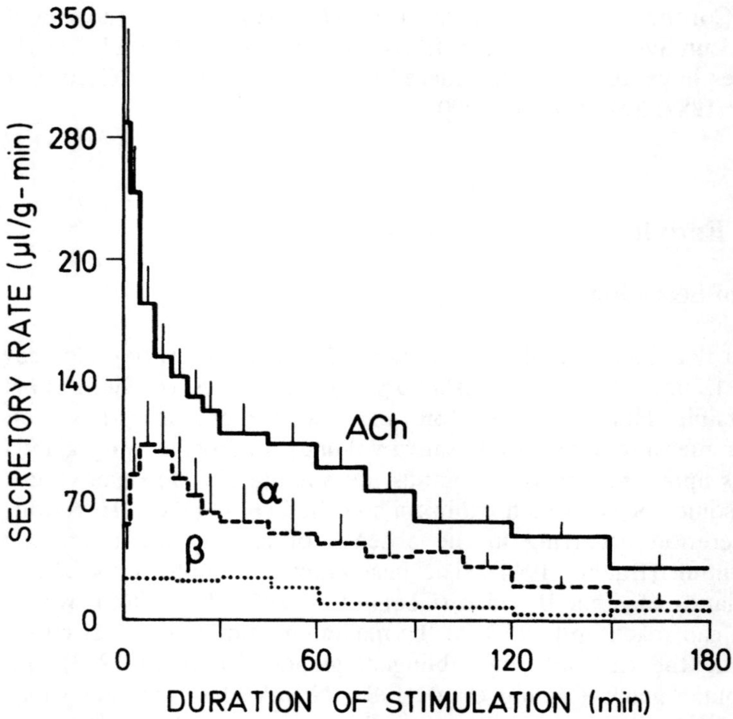

Fig. 1. Fluid secretion of rat mandibular glands in response to sustained stimulation with cholinergic or adrenergic agonists. Acetylcholine (ACh), phenylephrine (α) or isoproterenol (β) were infused at 10^{-6} mol/l. Atropine, propranolol or phentolamine (all 10^{-6} mol/l) were used to inhibit receptors other than the ones under study. (Novak 1990)

The salivary secretory rates are generally quite similar when corrected for gland weight. Thus, in mammals, for example, rat and rabbit mandibular glands can secrete up to 300 µl/g-min (Novak 1990), human submandibular and parotid glands secrete up to 200 µl/g-min. These rates are also comparable in marsupials: kangaroo and koala mandibular glands secrete maximally between 200 and 300 µl/g-min (Beal 1984, 1990a,b, 1991a). Except in sheep, the parotid glands usually secrete at lower rates than the mandibular glands (Bro-Rasmussen et al. 1956; Young and Schneyer 1981; Beal 1990a,b, 1991a,b). In comparison with most salivary glands, there are glands which secrete with exceptionally high rates that appear to be related to the specialized function of saliva in those species. For example, in dogs, in whom saliva has the function of cooling and regulating body temperature, secretion might rise tenfold above the rates of human saliva production (Adolph 1947). In ruminants, such as sheep and cows, whose digestion in the rumen requires large quantities of well-buffered fluid to neutralize the end product of fermentation, secretion might be higher by a factor of 4 (Kay 1960). The major salivary glands account for most of the total volume of saliva secreted, while the minor salivary glands account for most of the mucus secreted.

As will be discussed below, the salivary glands are often compared to the pancreas in their secretory mechanisms. Although the endpiece geometry is comparable in the two types of epithelia (Bolender 1974; Young et al. 1987), the maximal stimulated rates of secretion in the pancreas are much lower than in the salivary glands, e.g., 5–10 μl/g-min in the rat and rabbit pancreas (Bonting et al. 1980; Lingard and Young 1983; Seow et al. 1986, 1991) and a maximum of 20–40 μl/g-min for cat, pig and human pancreas (Case et al. 1980a; Grotmol et al. 1985; Go et al. 1986). Furthermore, in contrast to the salivary glands, the pancreatic acini are not surrounded by myoepithelial cells.

2.2 Composition of Primary Saliva

The electrolyte composition of the primary saliva was first determined by Martinez and coworkers (Holzgreve et al. 1966; Martinez et al. 1966) and Young and coworkers (Young and Schögel 1966; Young and Martin 1971) by micropuncture of the secretory/intercalated duct region of the intact rat mandibular gland. The collected primary saliva was isotonic and plasma-like, i.e., the concentration of Na^+ was about 140 mmol/l, K^+ was 7–8 mmol/l, and Cl^- was about 110 mmol/l. In the following years similar micropuncture studies were carried out on the mandibular glands of mouse, rabbit, cat and ferret (Mangos et al. 1973a,b, 1981; Kaladelfos and Young 1974; Endre and Young 1981), on the parotid glands of rat, mouse, rabbit, ferret and sheep (Mangos et al. 1966, 1973a,b, 1981; Mangos and McSherry 1969; Compton et al. 1980); and on the sublingual glands of rat and cat (Martin and Young 1971a; Kaladelfos and Young 1973). In general, for most glands it can be summarized that the unstimulated primary saliva has an osmolality between 290 and 310 mosmol/kg H_2O, the Na^+ concentration ranges from 130 to 160 mmol/l, and the Cl^- concentration is 100–130 mmol/l. The primary K^+ concentration is slightly above plasma values, ranging from 4 to 12 mmol/l. Except for the ferret salivary glands (Mangos et al. 1981), the primary concentration of HCO_3^- has not been directly measured. However, this can be inferred from the residual anion deficit (i.e., the summed number of cation equivalents minus anion equivalents), and the estimates are between 20 and 40 mmol/l for most glands (Young et al. 1980). These values might be even lower if one takes into account that the residual anions also probably include inorganic phosphate, which most glands secrete at concentrations 3 times those found in plasma. Interestingly, the concentration of inorganic phosphate in the sheep parotid gland was found to be about 10 times higher than in plasma, i.e., 12–14 mmol/l (Compton et al.

1980). With respect to other ions, this gland from sodium-replete animals is also an exception to the above described glands. The Cl^- concentration in primary saliva is lower than in other glands, 50–60 mmol/l, and the estimated HCO_3^- concentration could be as high as between 50 and 60 mmol/l (Compton et al. 1980). The K^+ concentration is about 10–12 mmol/l. Kangaroo and koala parotid glands, like the sheep parotid glands, produce high levels of phosphate and HCO_3^- in their final saliva (Beal 1984, 1990b). However, since these glands have not been micropunctured, the site of secretion of these anions is unknown (see also Sect. 2.3.1).

In most of the micropuncture studies described above, the effect of stimulation, parasympathomimetic and adrenergic, on the primary saliva composition has been investigated. In general, following stimulation, no effect on total osmolality, Na^+ or Cl^- was detected, but in some studies a small fall in K^+ was found (Martin and Young 1971a; Young and Martin 1971). Overall, the effects of stimulation on the primary fluid composition are small, and we can conclude that the most pronounced effect of stimulation must be on the rate of primary fluid secretion.

With some exceptions then, e.g., the sheep parotid, the micropuncture data confirm Thaysen's original postulate that the primary saliva resembles plasma and remains so during stimulation. We should, however, keep in mind that in all micropuncture studies, except in one study of the cat sublingual gland (Kaladelfos and Young 1973), samples have been obtained from the endpiece/intercalated duct region rather than the endpiece lumen, which is too narrow for micropuncture. Hence, it cannot be excluded that the intercalated ducts modify the endpiece secretion, perhaps by secreting themselves, as they are of the same embryological origin as the endpiece cells, and they are surrounded by myoepithelial cells (Young and van Lennep 1978). Perhaps it would be valuable with the new advanced techniques to micropuncture and microperfuse secretory endpieces and intercalated ducts separately (also see Sect. 3).

2.3 Electrolyte and Water Transport

In recent years it has been accepted that the transport of two major extracellular anions, Cl^- and HCO_3^-, could create the basic drive for salt and water transport in many secretory and absorptive epithelia. Lundberg proposed (as early as 1957) that the active transport of Cl^- was the prime driving force for salivary secretion. He based his theory on the observation in the cat sublingual gland that cholinergic stimulation caused the transepithelial potential of the secretory endpieces to become more negative and the basolateral membrane of secretory cells to hyperpolarize (Lundberg 1957c). This "secretory potential" could be due to activation of an electrogenic Cl^- pump or to a change in the membrane conductance to K^+, an event resulting in K^+ efflux. A transient increase in the K^+ concentration in the venous blood and saliva are in fact associated with the onset of cholinergic and α-adrenergic stimulation (Burgen 1956). Nevertheless, since the same hyperpolarization was observed when the membrane potential was raised above the K^+ equilibrium potential, Lundberg attributed the hyperpolarization to the current carried by the basolateral Cl^- pump. This theory was further supported by his observation that the secretory potential, as well as secretion itself, was considerably decreased when extracellular Cl^- was replaced by NO_3^- (Lundberg 1957c). His theory fell into disfavour, however, when a number of subsequent studies demonstrated that the secretory potentials were not always associated with secretion, and they were not always hyperpolarizing, but could be depolarizing or even biphasic (Petersen 1970a, 1980). These potentials were associated with an increase in the membrane permeability to K^+ and Na^+, resulting in a K^+ flux out of the cell and a Na^+ flux into the cell (Imai 1965; Petersen 1970a, 1972; Wakui and Nishiyama 1980a,b). Consequently, secretion was viewed in terms of the active transport of a cation rather than an anion, and a secretory model was proposed in which Na^+ was transported actively from the cytosol into the lumen by a membrane-bound ATPase (Petersen and Poulsen 1969; Petersen 1972; Imai 1976). This interest in Na^+ transport was sparkled by the studies of Ussing and Leaf (Koefoed-Johnsen and Ussing 1958; Leaf 1965) on "tight" epithelia (i.e., toad skin and amphibian bladder), where Na^+ transport is active, while Cl^- transport usually plays a passive role. In "leaky" epithelia, however, this model of ion transport is now found to be an exception rather than the rule. Nevertheless, in salivary glands no evidence for the luminal Na^+ pump was found, except for one study in which some $(Na^+ + K^+)$-ATPase activity was detected in the luminal membrane (Conteas et al. 1986).

The interest in Cl^- transport as the basic drive for secretion was rekindled when it was appreciated that the intracellular Cl^- concentration was far above the electrochemical equilibrium in many cells (Frizzell et al. 1979) and when it became accepted that transepithelial electrolyte transport does not need to involve primary active transport but can utilize ion gradients created by an active transport (Aronson 1981). Silva and coworkers (1977) proposed a coupled Na^+ and Cl^- transport (i.e., Na^+–Cl^- cotransporter) on the basolateral membrane of the shark rectal gland, which would utilize the Na^+ gradient created by the $(Na^+ + K^+)$-ATPase to drive Cl^- into the cell. Subsequently, Cl^- would enter the lumen through an apical Cl^- channel. Such a simple cotransporter, found for example in the renal early distal tubule, is sensitive to thiazides (Velázquez and Wright 1986; Greger and Velázquez 1987). Later, it was discovered that the Na^+–Cl^- cotransporter requires K^+ in Ehrlich ascites tumour cells (Geck et al. 1980) and that such a Na^+–$2Cl^-$–K^+ cotransporter is also present in absorptive and secretory epithelia (Greger et al. 1983; Greger and Schlatter 1984). This K^+-dependent cotransporter is sensitive to loop diuretics (Schlatter et al.

1983) and has now been proposed for many epithelia, including the salivary glands (see Sect. 2.3.1).

An alternative transport model has been proposed for many "leaky" epithelia to explain their ability to transport Na^+ and Cl^- independently of one another, and to explain the transport of HCO_3^- and the involvement of carbonic anhydrase and H^+ in salt and water transport. This model includes two independent antiports, one for Na^+/H^+ exchange and the other for Cl^-/HCO_3^- (OH^-) exchange. One or both of these exchangers can also be involved in volume and intracellular pH regulation. In the following section I will present the available evidence for Na^+–$2Cl^-$–K^+ cotransport and Na^+/H^+ (and Cl^-/HCO_3^-) exchangers and other components of the transport system [i.e., ($Na^+ + K^+$)-ATPase, K^+ and Cl^- channels] in salivary glands and describe to what extent they might be involved in the secretory mechanisms of various salivary glands.

2.3.1 Basolateral Membrane

2.3.1.1 Na^+-$2Cl^-$-K^+ Cotransporter

Some of the initial evidence for this cotransporter was obtained from whole gland studies of the cat sublingual and cat, rabbit and rat mandibular glands. These glands are relatively easy to perfuse in vivo and in vitro, and they secrete in response to cholinergic stimuli at a normal rate if they are provided with Cl^- as the major extracellular anion (i.e., in the absence of exogenous HCO_3^-), and the inhibition of carbonic anhydrase has little influence on secretion (Fig. 2) (Lundberg 1957c; Petersen 1971; Case et al. 1984; Novak and Young 1986; Novak et al. 1984). Salivary secretion, however, is inhibited by the removal of extracellular Na^+, Cl^- or K^+ (Lundberg 1958; Petersen and Poulsen 1967a; Petersen 1970b; Murakami et al. 1986). Shortly after the description of the Na^+-$2Cl^-$-K^+ cotransporter in other epithelia, Petersen argued that this mechanism is also responsible for Cl^- uptake in the salivary glands (Petersen and Maruyama 1984). He based his arguments on an observation that K^+ reuptake (following the initial acetylcholine-evoked release) is dependent on Cl^- in the medium, i.e., it occurs only if Cl^- (not NO_3^-) is present in the perfusion medium (Petersen 1970b). A similar anion preference had already been proposed by Lundberg for the cat sublingual gland, where only Cl^- and Br^- supported salivary secretion efficiently, whereas NO_3^-, I^- and SCN^- could only support 10–25% of maximal secretion (Lundberg 1958). A similar anion preference for secretion was found in the rabbit mandibular gland, where anions support secretion in the following sequence $Br^- \geqslant Cl^- > I^- \geqslant NO_3^- >$ isethionate (Novak and Young 1986). In the rat mandibular gland, however, Br^- was less efficient than Cl^- (Martinez and Cassity 1985a). Of course, in the intact gland the preference of the secretory system for anion will depend on

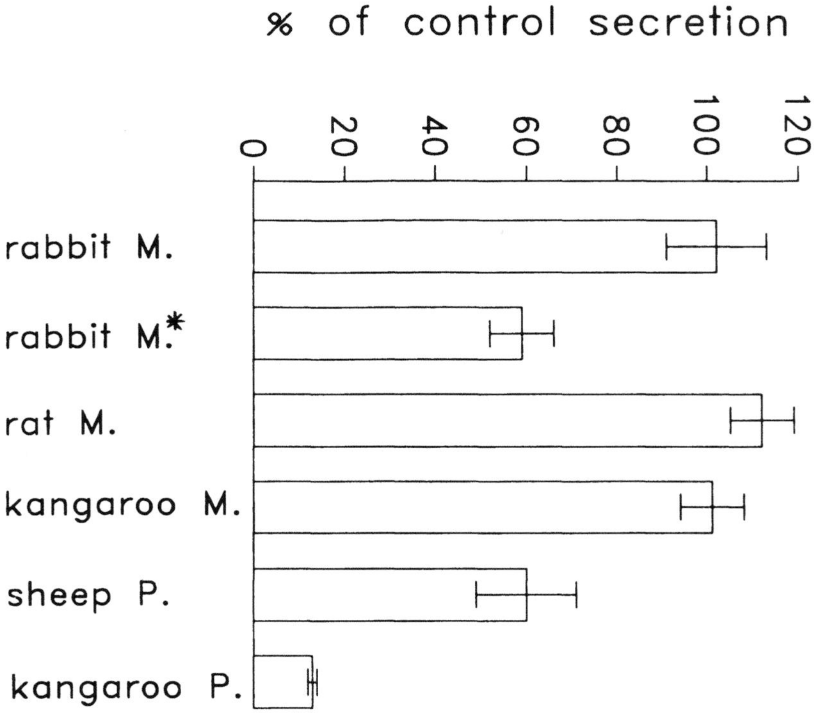

Fig. 2. Effect of carbonic anhydrase inhibition on various cholinergically stimulated salivary glands. Rat and rabbit mandibular (*M.*) glands were isolated and perfused in vitro. In rabbit glands secretion was evoked with 0.8 µmol/l or 80 µmol/l (*) acetylcholine and in rat glands, with 0.3 µmol/l acetylcholine. Perfusates contained acetazolamide or methazolamide (0.1–1 mmol/l). Sheep and kangaroo parotid (*P.*) glands were studied in vivo. In the sheep gland acetazolamide, methazolamide and benzolamide were used at concentrations of 1–10 mmol/l in carotid plasma. In the kangaroo acetazolamide was 0.5 mmol/l. Data were recalculated from Case et al. (1982); Pirani et al. (1987); Blair-West et al. (1980); Beal (1992)

the cotransporter, a Cl^-/HCO_3^- exchanger (if present) and luminal Cl^- channels, and which of these transport steps is rate-limiting in the overall Cl^- transport. Regarding the Cl^-/HCO_3^- antiport, we have shown that the rabbit and rat mandibular glands possess the antiporter (see below), but they can also secrete at a normal rate in the absence of HCO_3^- (Case et al. 1984; Novak et al. 1984). In that case they are insensitive to the Cl^-/HCO_3^- inhibitor SITS (4-acetamido-4′-isothiocyano-2,2′-stilbene disulphonic acid) (Novak and Young 1986). Furthermore, NO_3^- is transported quite well by the exchanger in the rabbit parotid gland (Turner and George 1988a) and in other cells (Warnock and Yee 1981; White and Imon 1983). Very little is known about the luminal Cl^- channels in the salivary glands (see Sect. 2.3.2). From studies in other epithelia, the anion selectivity of Cl^- channels (examined in excised patches), whole-cell currents and fluxes are a matter of debate at the moment. Overall, the anion preference revealed by experiments in perfused glands resembles those reported in a number of tissues where the Na^+-$2Cl^-$-K^+ symport has been identified (Geck et al. 1980; Chipperfield 1981; Greger et al. 1983; Turner and George 1988b).

The detailed sensitivity of the cotransporter to various loop diuretics is unknown in intact glands; however, furosemide at high doses of 0.1 and 1 mmol/l can inhibit 80% and 100% of the Cl^--supported secretion evoked by acetylcholine in the rabbit mandibular gland (Novak and Young 1986). In the rat mandibular gland 1 mmol/l furosemide inhibited 60–80% of secretion supported by Cl^- and HCO_3^- (Martinez and Cassity 1983; Novak et al. 1984; Pirani et al. 1987). Interestingly, furosemide also inhibited 60 and 90% of phenylephrine and isoproterenol stimulated secretion in the rat mandibular gland, respectively, and it caused an unexplained increase in the salivary Ca^{2+} concentration from about 5 to 15 mEq/l (Martinez and Cassity 1984). In dispersed acini of rat mandibular and parotid glands, the $^{36}Cl^-$ uptake is reduced by furosemide and in K^+-free media (Martinez and Cassity 1985b; Nauntofte and Poulsen 1986). The K^+ uptake in mouse mandibular acini depends on Na^+ and Cl^- and is abolished by piretanide (Exley et al. 1986). Measurements of O_2 uptake in the intact rabbit mandibular gland support the view that the acinar cotransporter has a stoichiometry of $1Na^+ : 2Cl^- : 1K^+$ (Smaje et al. 1986).

More direct evidence for the Na^+–$2Cl^-$–K^+ cotransporter was provided by Turner and coworkers (Turner and George 1988a,b) working with basolateral membrane vesicles from the rabbit parotid gland. The Na^+ uptake was markedly dependent on both K^+ and Cl^- in the extravesicular medium, and the data are consistent with the stoichiometry of $1Na^+ : 2Cl^- : 1K^+$. The half-saturation constant for Cl^- of about 90 mmol/l is somewhat higher than that seen in reabsorptive epithelia, and furosemide inhibited this cotransporter with a relatively high $K_{0.5}$ of 2×10^{-4} mol/l. The authors point out that this may be due to the non-physiological ionic concentrations and room temperatures used in this study. The parotid acini seem to be a rich source of the cotransporter (Turner and George 1988b), and studies are currently under way to purify the cotransporter using a more potent diuretic, bumetanide, to monitor its binding properties (Turner and George 1990). The bumetanide binding protein-detergent-lipid complex is about 200 kDa. In their recent study (Corcelli and Turner 1991), the authors report that the affinity of bumetanide for protein depends on certain minor phospholipids, which may be the constituents of the plasma membrane in situ.

2.3.1.2 Na^+/H^+ and Cl^-/HCO_3^- Exchangers

From studies on isolated rat and rabbit mandibular glands, it became apparent that the transport of Cl^- is the driving force for secretion and that the Na^+-$2Cl^-$-K^+ cotransporter accounts for most of the secretion. Inhibition of carbonic anhydrase has little effect on secretion, unless the secretion is evoked by supramaximal doses of acetylcholine (Fig. 2), in which case the inhibitor decreases the HCO_3^- concentration in saliva by about 10 mmol/l (Case et al. 1982; Pirani et al. 1987). Nevertheless, these glands also have a capacity to drive secretion utilizing HCO_3^- transport. This phenomenon is best unmasked if the Cl^- transport is prevented by furosemide or simply if Cl^- is removed from the perfusate, and HCO_3^- is left as the only other transportable anion. Under such conditions rat and rabbit mandibular glands can secrete at 30–40% of their normal rate (Case et al. 1984; Novak et al. 1984; Novak and Young 1986; Pirani et al. 1987). This secretion, in contrast to normal or Cl^--supported secretion, is rich in HCO_3^- (hence the acini must secrete HCO_3^-). Furthermore, this secretion is sensitive to amiloride and carbonic anhydrase inhibition, and as a result we postulated that the basolateral membrane also possesses a Na^+/H^+ antiport which is involved in HCO_3^- secretion. It is also worth noting that the ability of HCO_3^-/CO_2 to support secretion is about the same as that of Cl^- or acetate, if these anions are present in the perfusate in the same concentration, i.e., 25 mmol/l (Fig. 3; Novak 1984; Novak and Young 1986). Under physiological conditions, i.e., HCO_3^-/Cl^- perfusion, such HCO_3^- transport in the mandibular glands would account for 20–30% secretion, the rest being due to Cl^- transport via the Na^+-$2Cl^-$-K^+ cotransporter (Novak and Young 1986; Pirani et al. 1987).

The contribution of the Cl^- as opposed to the HCO_3^- transport mechanism responsible for the formation of primary saliva, however, may not be the same in all salivary glands. For example, the parotid glands of foregut fermenters appear to be more avid HCO_3^- transporters. In the sheep parotid gland in vivo, furosemide and amiloride have no effect on secretion (Wright et al. 1986), but carbonic anhydrase inhibitors infused into the arterial supply have small but significant effects on basal and parasympathetically stimulated secretion, reducing it by about 20 and 40%, respectively (Fig. 2; Blair-West et al. 1980). At the same time the saliva remains quite rich in HCO_3^- (i.e., $\approx$100 mol/l) and is affected only by 10% by the inhibitors, while the phosphate and Cl^- concentrations rise correspondingly. Interestingly, the kangaroo parotid gland seems to be similar to that of sheep in many respects, and the HCO_3^- transport is even more pronounced. The Cl^- transport inhibitors, furosemide and SITS, had no effect on secretion at millimolar doses in carotid plasma (Beal 1990a). Amiloride had a dose-dependent effect on secretion, e.g., at 0.5 mmol/l the stimulated secretory rate decreased by $\approx$90%, and the HCO_3^- level decreased, while Cl^- and phosphate correspondingly increased (Beal 1990a). Most striking are the effects of carbonic anhydrase inhibition. Following acetazolamide (0.5 mmol/l) infusion into the artery supplying the gland, acetylcholine-stimulated secretion decreased by about 87%, the HCO_3^- concentration fell from 130 to 80 mmol/l (taking into account the

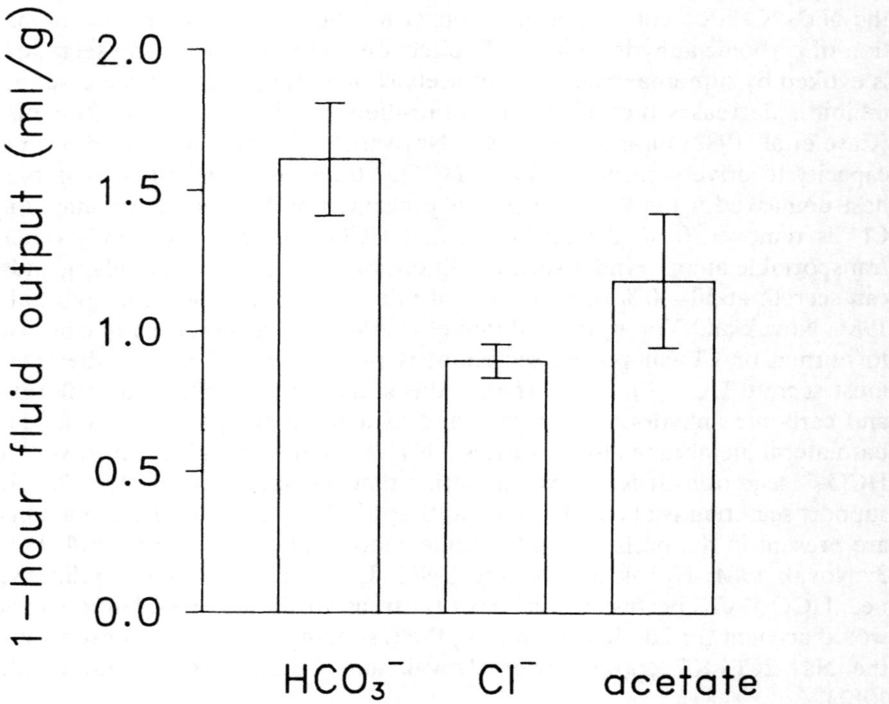

Fig. 3. Efficiency of various anions (25 mmol/l) in supporting acetylcholine (0.8 µmol/l) stimulated secretion in the perfused rabbit mandibular glands (*n* = 4–6). The remaining anion in the perfusate was isethionate (121 mmol/l). For comparison, the 1-h fluid output in glands perfused with the control solution (25 mmol/l HCO_3^- and 121 mmol/l Cl^-) was 6.2 ± 0.5 ml/g tissue weight (*n* = 20). (Data from Novak 1984)

lower secretory rate), and concurrently the salivary Cl^- rose from 6 to 28 mmol/l, and phosphate rose from 14 to 28 mmol/l (Beal 1990a, 1991b). These changes in ionic concentrations are greater than could be accounted for by the lower secretory rates, i.e., ductal modification (see Sect. 3). Rather, it would seem that the gland has changed from a HCO_3^- secretor to a Cl^-/phosphate secretor. Most interestingly, the superimposition of isoprenaline on acetylcholine revived the carbonic anhydrase inhibited secretion to about 40% of the control rate, but this was not accompanied by an increase in the salivary HCO_3^- concentration. In contrast to the kangaroo parotid gland, the mandibular gland of the same animal is similar to mandibular glands of other species in its relatively low salivary HCO_3^- concentrations and low sensitivity to carbonic anhydrase inhibition (Beal 1991b; Fig. 2).

The findings in sheep and kangaroo parotid glands would be consistent with the idea that HCO_3^- transport is the driving force for some fluid transport. The question remains, where is the site of this electrolyte and fluid secretion? The acinar-intercalated duct fluid from the sheep parotid gland is not particularly rich in HCO_3^- (estimated from anion deficit) (Compton et al. 1980). However, some ductal modification could have taken place prior to the collection site. Alternatively, more distal ducts could be the site of HCO_3^- secretion, and some fluid secretion must follow. Beal (1991b), however, argues that excurrent ducts in parotid glands from Na^+-replete kangaroos form less than 2% of the total gland volume; thus, their contribution to HCO_3^- transport would most likely be small. It is also quite possible that the HCO_3^- and phosphate/Cl^- transports are inversely linked in the two glands. In the sheep parotid gland, parathyroid hormone, probably acting only on the endpieces, increases the salivary phosphate and fluid secretion, while the salivary HCO_3^- concentration is lowered (Wright et al. 1984). In agreement with this interpretation is the recent finding of Vayro and coworkers (1991) who found a Na^+-dependent phosphate carrier on the basolateral plasma membrane vesicles of sheep parotid acini.

More direct evidence for the Na^+/H^+ antiport in the parotid glands is provided by flux and pH studies of rat parotid acini, although these glands achieve lower HCO_3^- concentrations in their saliva, i.e., 20–40 mmol/l (Young and Schneyer 1981). In rat parotid acini suspended in HCO_3^-/Cl^- media and stimulated with carbachol, 75% of the total Na^+ influx consisted of the dimethylamiloride-sensitive component (i.e., Na^+/H^+ exchange), while only ≈15% consisted of the bumetanide-sensitive cotransport (Dissing and Nauntofte 1990). In a similar preparation Soltoff et al. (1989) measured only 20% of amiloride-sensitive Na^+ flux, but the media used were free of HCO_3^-. In a similar preparation of rat parotid acini, muscarinic and α-adrenergic stimulation induced a rapid increase in Na^+/H^+ activity, reflected by an increase in intracellular pH (pH_i) (Melvin et al. 1988). However, if amiloride was added to the preparation, carbachol induced dramatic acidification, presumably unmasking a rapid loss of HCO_3^- that cannot be balanced by H^+ loss via the Na^+/H^+ exchanger (Melvin et al. 1988; Soltoff et al. 1989). Nauntofte and Dissing (1988a) obtained a resting pH_i of ≈7.37 which upon cholinergic stimulation transiently decreased by about 0.1 unit (10 s). This pH_i recovered within 1 min, unless inhibited with amiloride. In the intact rabbit mandibular gland and in isolated acini, acetylcholine caused a dose-dependent transient pH_i decrease followed by recovery and alkalinization to a new steady-state (Lau et al. 1989; Steward et al. 1989). In the perfused rat mandibular gland, however, only a stimulation-induced fall in pH_i was observed (Pirani et al. 1987), while in the rat mandibular acini the stimulation evoked the usual small transient acidification followed by sustained alkalinization (Okada et al. 1991). In any given cell, the intracellular pH would depend on the stimulation-induced HCO_3^- efflux (see Sect. 2.3.2) and activation of the Na^+/H^+ exchanger. Activation of the Na^+/H^+ exchange is not associated with protein kinase C (Manganel and Turner 1990, 1991; Okada et al. 1991), but it is dependent on Ca^{2+}

(Nauntofte and Dissing 1988a; Manganel and Turner 1990; Okada et al. 1991). In addition, it is postulated that there is also a fast component of activation of Na^+/H^+ exchange, which may be an osmoregulatory upregulation following the stimulation-induced cell shrinkage (Manganel and Turner 1991). The cells loose about 30% of their K^+, 50% of Cl^-, 20% of HCO_3^- and 50% of water (Soltoff et al. 1989; Manganel and Turner 1991). Overall then, Na^+/H^+ exchange is important in maintaining both pH_i and ionic homeostasis during fluid secretion and in providing HCO_3^- for transport across the luminal membrane.

Returning to the intact glands, we have observed in rabbit mandibular glands that disulphonic stilbenes, e.g., SITS, had an influence on HCO_3^-/Cl^--supported secretion; paradoxically, it would seem that secretion was increased by about 30% in paired experiments (Novak and Young 1986). A similar trend was found in the rat mandibular gland (Pirani et al. 1987). We have concluded that this was due to the basolateral Cl^-/HCO_3^- antiport, the inhibition of which would lead to an increase in pH_i (Pirani et al. 1987; Lau et al. 1989), which is known to stimulate salivary secretion (Novak et al. 1980). Furthermore, there might be an increase in the HCO_3^- secretion as there was a slight, but significant increase in the salivary HCO_3^- and corresponding decrease in the Cl^- concentration (Novak 1984; Novak and Young 1986). In contrast to our studies, Lau and coworkers (1990) found no effect of another disulphonic stilbene, DIDS (4,4′-diisothiocyanostilbene-2,2′-disulphonic acid) (0.1 mmol/l), on the acetylcholine-stimulated secretion of the rabbit mandibular glands; however, in unstimulated glands DIDS (1 mmol/l) alone evoked secretion. Nevertheless, from the DIDS sensitivity of pH_i in the rabbit mandibular acini, they conclude that the Cl^-/HCO_3^- exchanger is present (Lau et al. 1989). From intact gland studies it is difficult to deduce the sidedness of the inhibitor action, although in the case of SITS it is unlikely that at a 1–5 μmol/l concentration, which reaches the lumen, the drug could have inhibited transport processes on the luminal membrane (Novak 1984; Novak and Young 1986). Again, clearer on this issue are the studies on basolateral membrane vesicles isolated from rat parotid acini, where SITS-sensitive and Na^+-independent Cl^-/HCO_3^- was detected (Turner and George 1988a). This is supported by the flux studies on isolated parotid acini suspended in a HCO_3^- medium. The Cl^- reuptake into the acini, following a carbachol-induced K^+ and Cl^- loss, occurs at the same rate in low and high Na^+ medium, suggesting that the Cl^-/HCO_3^- exchanger is responsible for keeping Cl^- above equilibrium under physiological conditions (Dissing and Nauntofte 1990).

2.3.1.3 (Na^+ + K^+)-ATPase and Basolateral K^+ Channels

The prime driving force for secretion (Fig. 4) is the basolaterally located (Na^+ + K^+)-ATPase. The pump actively transports 3 Na^+ out of and 2 K^+ into the cell, thereby creating an inwardly directed gradient for Na^+ and an

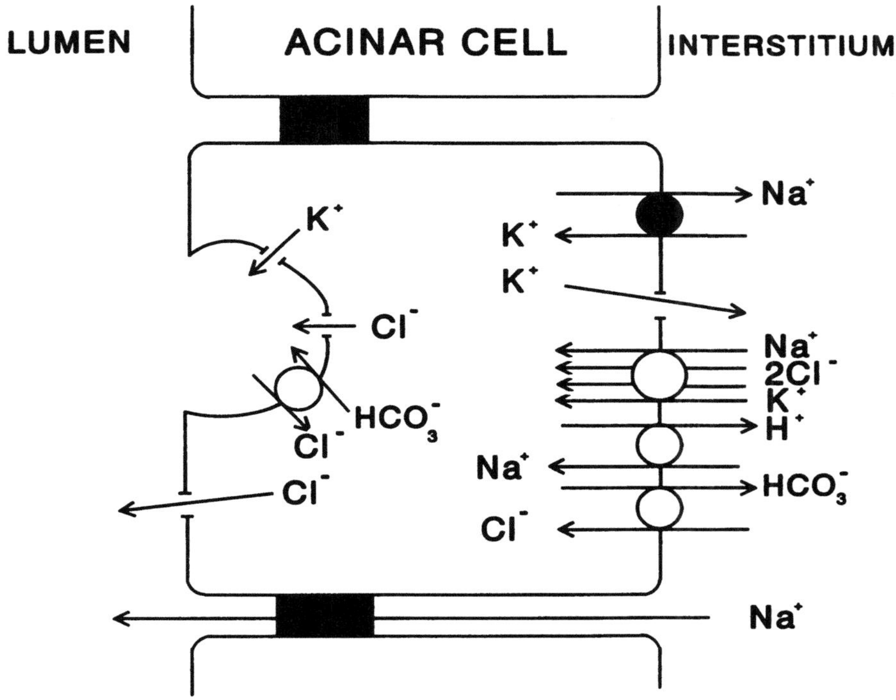

Fig. 4. Composite model for electrolyte transport in a secretory endpiece cell. Different carriers on the basolateral membrane can account for Cl^--driven secretion via the Na^+–$2Cl^-$–K^+ symport, HCO_3^--dependent Cl^- secretion via the Cl^-/HCO_3^- antiport and HCO_3^- secretion via the Na^+/H^+ antiport. The predominance of one of these transport systems over another may account for some differences between salivary glands, e.g., Cl^- secretors such as rodent mandibular glands and HCO_3^- secretors such as kangaroo parotid glands. The driving force for ion flow is created by the (Na^+ + K^+)-ATPase. The prime steps in initiating cholinergic secretion include opening of basolateral K^+ channels and postulated luminal Cl^- channels. The properties of the luminal membrane are largely unknown. Secretory granules contain Cl^- and K^+ conductances and a Cl^-/HCO_3^- antiport. It is postulated that exocytotic fusion of the granule with the membrane might account for protein and some electrolyte and fluid secretion. For details, see the text

outward one for K^+. The chemical gradient for Na^+ is then utilized by the cotransporter to deliver Na^+, Cl^- and K^+ into the cell. The electrogenicity of the pump in mammalian gland cells has been demonstrated in a number of electrophysiological studies summarized by Petersen (1987). In addition, the (Na^+ + K^+)-ATPase in salivary glands has been demonstrated morphologically and functionally. For example, Bundgaard and coworkers (1977) found the ATPase on the basolateral membrane of cat sublingual and mandibular glands. Conteas and coworkers (1986) showed the basolateral (and some luminal) localization of the ATPase in rat parotid acini. Interestingly, recent histochemical studies of Koob et al. (1990) show that the (Na^+

$+ K^+$)-ATPase and a cytoskeletal protein ankyrin are colocalized on the basolateral membrane of pig parotid acini and striated ducts. It is believed that such a connection to the cytoskeleton would capture integral proteins, like the ATPase, at a specialized site of the cell surface and prevent it from clustering and undergoing endocytosis.

Functionally, it has been shown that the removal of extracellular K^+, which would affect the pump (and the cotransporter), also inhibits secretion (Petersen and Poulsen 1967b; Laugesen et al. 1976). In many salivary glands ouabain inhibits secretion evoked by various agonists including acetylcholine, α- and β-adrenergic agonists and ATP (Petersen and Poulsen 1967b; Compton et al. 1981; Martinez and Cassity 1983, 1984; Evans et al. 1986; Pirani et al. 1987). There is evidence that the stimulation of secretion might actually involve activation of the pump, so that the Na^+ and K^+ gradients are re-established after stimulation (Stewart et al. 1983; Soltoff et al. 1989; Dissing and Nauntofte 1990).

Potassium accumulated in the cell by the ($Na^+ + K^+$)-ATPase and the Na^+-$2Cl^-$-K^+ cotransporter leaves the cell via the basolateral K^+ channels (Fig. 4). The resting membrane potential of the salivary secretory cells is between −70 and −60 mV, indicating that it is due to a large extent to the K^+ conductance (Nishiyama and Petersen 1974; Roberts et al. 1978; Wakui and Nishiyama 1980a,b; Gallacher 1982). The opening of the K^+ channels is a leading membrane transport event in the stimulation of salivary secretion (see below) and explains the long-standing observation of K^+ transients, i.e., an increase in venous K^+ concentration shortly after cholinergic stimulation (Burgen 1956). It has now been shown in many salivary glands that stimulation with acetylcholine, α-adrenergic agonists, substance P and ATP induces a K^+ flux out of the cell (Putney 1976; Poulsen and Bledsoe 1978; Katoh et al. 1983, 1986; Petersen 1987; Nauntofte and Dissing 1988b; Soltoff et al. 1990). Correspondingly, under certain conditions a stimulation-induced increase in the basolateral membrane conductance to K^+ has been detected as an initial hyperpolarization of the membrane potential in single secretory cells in several glands (Lundberg 1957b, 1958; Nishiyama and Petersen 1974; Roberts et al. 1978; Wakui and Nishiyama 1980a,b). However, the pattern of membrane potential changes following stimulation can be quite complex, not only hyperpolarizing, but also depolarizing or biphasic, where the response depends on the initial resting membrane potential and the dose of agonist used for stimulation (Roberts and Petersen 1978; Roberts et al. 1978; Wakui and Nishiyama 1980a,b). Wakui and Nishiyama (1980a,b) inferred from their microelectrode studies on mouse mandibular acini that there are two types of cation channels on the basolateral membrane, one sensitive to tetraethylammonium (TEA^+) and responsible for the K^+-dependent hyperpolarization, and the other responsible for the Na^+-dependent membrane depolarization. Since the hyperpolarizing potential changes induced by acetylcholine do not reach the K^+ equilibrium potential, unless Cl^- is replaced by sulphate, the activation of a Cl^- conductance has been implicated (Nishiyama and Petersen 1974; Roberts et al. 1978). This Cl^- conductance, however, is most likely located on the luminal membrane (see Sect. 2.3.2). The initial membrane potential response is often followed by a delayed hyperpolarization which cannot be explained as an increase in the K^+ conductance. Rather, it is thought to be due to activation of the electrogenic (Na^+ + K^+)-ATPase, since it is blocked by ouabain or exposure to Na^+ and K^+-free media (Roberts et al. 1978).

Supporting evidence for K^+ channels has been obtained in patch-clamp studies. Several types of channels have been described on cell-excised patches of the basolateral membrane (see Petersen 1987; Petersen and Gallacher 1988; Cook and Young 1990 for reviews). The first type is a Ca^{2+}-sensitive "maxi" K^+ channel with a conductance of ≈200 pS (in symmetrical KCl solutions); it is highly selective for K^+, even discriminating against Rb^+, and it is inhibited by Ba^{2+} and TEA^+ (Maruyama et al. 1983; Gallacher et al. 1984; Petersen 1986). The channel is activated by increases in the cytosolic Ca^{2+} and by membrane depolarization (Maruyama et al. 1983; Petersen and Maruyama 1984). Two other types of K^+ channels have been seen in rodent mandibular acini and sheep parotid acini: small conductance 30–50 pS channels and an intermediate conductance ≈80 pS channel, which might be a substate of the maxi K^+ channels (Maruyama et al. 1983; Cook et al. 1988a, 1990b; Wegman et al. 1988). However, the detailed properties of these channels in salivary glands are unkown. Another well-characterized type of channel only found in cell-excised patches is a so-called Ca^{2+} activated non-selective cation channel, which has a conductance of 30–35 pS under quasiphysiological conditions; it cannot distinguish between Na^+, K^+, Li^+ and Rb^+, but it is impermeable to anions (Maruyama et al. 1983; Gallacher et al. 1984). This latter channel, but not the maxi K^+ channel, is also found in rodent pancreatic acini (Petersen and Gallacher 1988). The possible role of this channel in secretion is unclear, but various hypothetical models have been described (Petersen 1986, 1987; Cook et al. 1990a; Soltoff et al. 1990).

In order to verify whether or not any given K^+ channel is responsible for the permeability changes associated with secretagogue stimulation, supporting evidence is required from intact cells. One method is to characterize the single channels on the cell. Various agonists, utilizing Ca^{2+} as the intracellular messenger, have been shown to activate the Ca^{2+}-sensitive maxi K^+ channel in situ (Gallacher and Morris 1986, 1987; Morris et al. 1987). Acetylcholine increases the frequency and duration of opening of these K^+ channels, an effect mimicked by the Ca^{2+} ionophore A23187 and sustained depolarization in the presence of extracellular Ca^{2+} (Maruyama et al. 1983; Gallacher and Morris 1987). Under near physiological conditions the channel has a conductance of about 35 pS (Gallacher and Morris 1986). From conventional microelectrode recordings (see above) it seems quite clear that the resting membrane potential is dominated by the potassium conductance, but the open-state probability of the Ca^{2+}-sensitive K^+ channel is very low (Gallacher and Morris 1986), and it is estimated that there would be less than one channel open per cell (Petersen and Gallacher 1988), although the estimates differ (Cook and Young 1990). If the number of channels were less than one, this would obviously be a problem for a single cell, but one should consider that in the intact, unstimulated acinus each cell is coupled with 100–500 other cells. Upon stimulation the number of open channels per cell would increase to about 70, and interestingly the cells become less coupled (Petersen 1987). For the time being, these estimates from patch-clamp studies are at variance with K^+ flux studies (Nauntofte and Dissing 1988b), where the flux through the maxi K^+ channels was monitored as a carybdotoxin-sensitive component. From the data in these studies, it is estimated that at the resting membrane potential there are about 40 open channels per cell, and this increases to 3000 channels upon stimulation.

Another way to test the physiological relevance of a particular channel is to compare its characteristics to those of whole-cell voltage-clamp currents, which then should comprise many small elementary currents. Thus, the stimulation of acinar cells by various agonists that increase the cytosolic Ca^{2+} concentration and activate the phosphoinositol pathways results in a marked increase in the voltage-activated outward K^+ currents (Iwatsuki et al. 1985; Cook et al. 1988b). However, although in pig pancreatic acinar cells and lacrimal cells, TEA^+ in the extracellular medium blocks these acetylcholine-induced currents (Petersen 1987; Petersen and Gallacher 1988), in rodent parotid acini the effects of TEA^+ are controversial (Iwatsuki et al. 1985; Shigetomi et al. 1991). This finding suggests that acetylcholine may activate K^+ channels other than the classical maxi K^+ channels. Similar conclusion can be reached from the work of Soltoff and coworkers (1990) who measured the rate of initial K^+ efflux in rat parotid acini. Various blockers of the maxi K^+ channel, e.g., TEA^+ and carybdotoxin, had partial effects and inhibited K^+ efflux to a different extent depending on the type of agonist used, such as carbachol and ATP. Thus, multiple types of K^+-permeable channels might contribute to the agonist-induced K^+ loss. From studies on intact perfused rat and rabbit mandibular glands, Ba^{2+} and TEA^+ inhibit about 80–90% of acetylcholine-evoked secretion (Evans et al. 1986; Lau et al. 1990).

The sheep parotid gland has an unusually high spontaneous secretion (see Sect. 2.1) which is not sensitive to TEA^+. This has been linked to a low conductance K^+ channel (30 pS) which occurs at high density and with high open probability in cell-attached patches on parotid acini (Cook et al. 1990b). Acetylcholine stimulation has little influence on this channel; instead, it activates the Ca^{2+}- and voltage-sensitive K^+ channel in cell-attached patches. In contrast, the sheep mandibular gland does not secrete spontaneously, and the predominant K^+ channel is the maxi K^+ channel, which seems to have a different cation selectivity and sensitivity to TEA^+ and Ca^{2+} than the classical maxi K^+ channel (Cook et al. 1990b).

Taken together, the evidence described above suggests that the Ca^{2+}-activated maxi K^+ channels play a role in secretion evoked by cholinergic, α-adrenergic agonists and substance P, but most likely they are not the only channels involved. When it comes to β-adrenergic stimulation, the issue of K^+ channels and activation requires an extra word. Fluid secretion is generally small in response to β-agonists (see Sect. 2.1). However, protein secretion is quite prominent and associated with the activation of adenylate cyclase (Young et al. 1987). Interestingly, isoproterenol evokes K^+ uptake without a preceding release of K^+, which is also dependent on Na^+ and Cl^- and inhibited by furosemide and ouabain (Katoh et al. 1983, 1986). In contrast to this is the finding that isoproterenol increased the Rb^+ efflux (Putney et al. 1977). Note, however, that Rb^+ may not be a representative measure of K^+ efflux, at least for Ca^{2+}-sensitive maxi K^+ channels (Gallacher et al. 1984). Looking at membrane potentials, isoproterenol at low doses induces hyperpolarization without any change in the membrane input resistance, while at high doses the agonist induces depolarization or a biphasic response with a small change in input resistances (Roberts and Petersen 1978; Iwatsuki and Nishiyama 1982; Katoh et al. 1983, 1986). These findings would imply that isoproterenol stimulation might have little primary effect on the basolateral K^+ channels, and K^+ uptake has been ascribed to the (Na^+ + K^+)-ATPase and the Na^+-$2Cl^-$-K^+ cotransporter (Katoh et al. 1983, 1986). Nevertheless, recent findings of Cook and coworkers (1988a) using whole-cell recordings from rat mandibular acini show another interesting aspect of β-adrenergic stimulation. Isoproterenol, like acetylcholine, induces K^+ and Cl^- currents (currents measured at clamp potentials of 0 and $-80\,mV$) in whole-cell recordings of rat mandibular glands. These effects are not mimicked by cAMP, forskolin or cholera toxin but rather are dependent on an intracellular Ca^{2+} elevation. Thus, it appears that Ca^{2+}, not cAMP, mediates the K^+ and Cl^- currents in rat mandibular acini, and fluid secretion might come about via the same mechanism as in cholinergically stimulated glands. Nevertheless, since the cAMP concentration (and Ca^{2+} flux) does rise after isoproterenol stimulation and cAMP mediates protein secretion (Putney et al. 1977; Dreux et al. 1986), these findings cannot exclude the possibility that cAMP regulates secretion at some other steps than those involving K^+ and Cl^- conductances.

2.3.2 Luminal Membrane

Intracellular microelectrode studies indicate that the acinar cell membranes are permeable to Cl^- (Petersen 1980). Since the Cl^- conductance of the basolateral membrane is rather small (Roberts et al. 1978), and no basolateral Cl^- channels have been found by the patch-clamp technique, it is presumed that the Cl^- channels are located on the luminal membrane (Fig. 4). The luminal membrane of salivary secretory cells is rather inaccessible, and as a consequence its detailed properties are unknown, and parallels are often drawn from other Cl^--secreting epithelia. Here, I shall try to concentrate on what is actually known about salivary glands.

There are two studies using conventional electrophysiological techniques in which microelectrodes were placed into the lumen of secretory endpieces. Lundberg (1957a,b) found that the resting cat sublingual gland had a transepithelial potential of between 0 and +2 mV (lumen positive), and it became about −25 mV upon cholinergic stimulation. Most of this change was due to a 25 mV hyperpolarization of the basolateral membrane, but there was also a delayed hyperpolarization of the luminal membrane of about 10 mV. Similarly, Imai (1965) found in the dog mandibular gland that cholinergic stimulation increased the transepithelial electrical potential from −2 to about −12 mV, while the basolateral membrane potential hyperpolarized by about 15 mV; hence, the luminal membrane must have hyperpolarized by 5 mV. Using double-barrelled microelectrodes Lundberg (Lundberg 1957a) found that secretory cells had a resistance of 2 MΩ, which decreased to 1.8 MΩ after stimulation. Measuring resistance changes between the lumen and external medium, he ascribed a large part of this decrease to a decrease in the basolateral membrane resistance from 18 to 9 Ωcm^2, while the resistance of the luminal membrane decreased from 11 to 9 Ωcm^2. Both potential and resistance changes in the basolateral membrane are at present interpreted as due to opening of basolateral K^+ channels. The hyperpolarization of the luminal membrane remains to be confirmed and explained.

Due to inaccessibility of the luminal membrane in the intact gland and isolated acini, there is little direct information about single channel properties. In a preliminary report on rat mandibular endpieces, Ca^{2+}-sensitive Cl^- channels with a conductance of 14 pS (symmetrical CsCl solutions) were described (Martin et al. 1990). However, it is not clear on which membrane this channel is localized. In primary cultures or permanent cultures of acini, patch-clamp studies have failed to show single Cl^- channels. However, if the Cl^- channels had a conductance smaller than 5 pS, they may have gone undetected. It seems at the moment that channels which might be relevant for secretion are 4–8 pS in conductance (Gray et al. 1988, 1990; Tabcharani et al. 1990), or in fact they may be smaller than the resolution of the conventional patch-clamp technique (Hansen et al. 1991). Nevertheless, there is evidence for Cl^- currents from whole-cell current recordings from salivary tissue, although more is known about the pancreatic acini and lacrimal glands (Petersen 1987). In rat and mouse parotid acini and rat mandibular acini, cholinergic and β-adrenergic stimulation increases the inward flowing current, due to Cl^- efflux, an event which requires Ca^{2+} (Iwatsuki et al. 1985; Cook et al. 1988a,b; Shigetomi et al. 1991). Notably, the Cl^- current in the rat mandibular acini is larger and more sustained than in the lacrimal gland or pancreas (Findlay and Petersen 1985; Osipchuk et al. 1990; Petersen et al. 1991). Furthermore, in pancreatic acini various agonists, e.g., acetylcholine and cholecystokinin (CCK), induce cytoplasmic Ca^{2+} oscillations which are followed by synchronous, Ca^{2+}-dependent Cl^- currents (Osipchuk et al. 1990; Petersen and Wakui 1990; Petersen et al. 1991). In rat parotid acini similar temperature-dependent Ca^{2+} and whole-cell currents were detected (Gray 1988). In a recent study on the same cells, however, the Ca^{2+} oscillations observed are much slower than in pancreatic acini and followed by oscillations in intracellular Na^+ concentration and cell volume (Wong and Foskett 1991).

Upon cholinergic stimulation Cl^- leaves the cell, and the cell shrinks (Martinez and Cassity 1985b; Nauntofte and Poulsen 1986; Lau and Case 1988; Foskett 1990; Soltoff et al. 1990; Nakahari et al. 1991). The studies in which Cl^- channel blockers (Wangemann et al. 1986) were used in the hope of inhibiting the Cl^- channel are relatively inconclusive (Melvin et al. 1988; Braun et al. 1989), since at millimolar concentrations they are decouplers of oxidative phosphorylation (Lukacs et al. 1991) and can block other channels, e.g., non-selective cation channels (Poronnik et al. 1991).

Apart from secreting Cl^-, we have already seen that salivary glands can secrete HCO_3^-, but whether this takes place via the same postulated channels as for Cl^- is unknown. Studies on secretory tissues so far have shown that the selectivity of Cl^- channels for HCO_3^- is rather low under normal physiological conditions (Tabcharani et al. 1989; Gray et al. 1990; Kunzelmann et al. 1991). If, however, the extracellular Cl^- is removed, one could speculate that HCO_3^-, and also other anions like acetate, would take this route (Novak and Young 1986, 1989; Lee and Turner 1991) (Fig. 3). An attractive model which would incorporate HCO_3^- and Cl^- secretion is that proposed for pancreatic ducts, in which the luminal membrane contains Cl^- channels and Cl^-/HCO_3^- antiports in parallel, and full or partial recirculation of Cl^- would lead to a corresponding HCO_3^- secretion into the lumen (Novak and Greger 1988a). In fact, such a system has been proposed for the secretory granules (see Sect. 2.3.3). Without access to the luminal membrane, such a model for salivary endpieces, however, remains speculation.

Apart from K^+ channels on the basolateral membrane, K^+ channels on the luminal membrane have been considered in an equivalent circuit model (Cook and Young 1989a). Theoretically, these channels may not impede secretion and might even explain the following observations: initial loss of K^+ into saliva following stimulation (Burgen 1956), higher concentrations of K^+ in primary saliva than in plasma (see Sect. 2.2) and hyperpolarization of the luminal membrane observed by Lundberg (1957b) and Imai (1965). However, direct evidence for such channels is outstanding. As mentioned above, in cultured secretory epithelial cell lines, a 25 pS non-selective cation channel is found, presumably on the luminal membrane, if the epithelium is polarized (Cook et al. 1990a). Whether this channel plays a significant physiological role in secretion is speculative since it is inactive on the cell. It might allow Ca^{2+} entry into the cell (Poronnik et al. 1991), or more likely the entry of Na^+, which at present is not reconcilable with the current model for secretion. If such a channel were incorporated into the basolateral or luminal membrane of the secretory cell, the Na^+ shunt would force us to review the models for the secretion of electrolytes (Petersen 1987; Cook et al. 1990a), or as Ussing and Eskesen (1989) suggest, Na^+ recycling may be a necessary part of isotonic water transport.

2.3.3 Secretory Granules

One of the most striking features of secretory cells within an endpiece is that the apical pole of the cytoplasm is densely packed with secretory granules. In this section I would like to consider secretory granules not from the point of view of their macromolecular contents, but rather from that of their possible link to fluid secretion. It has been known for many years that anions are important for the exocytosis and lysis of, for example, neurosecretory granules, adrenal chromaffin and neutrophil secretory granules. In the past 5 years, it has become apparent that some properties of the luminal membrane of secretory exocrine cells we are searching might actually reside in the membrane of secretory granules. Hence, the finding of a Cl^- conductance in isolated pancreatic zymogen granules initiated the hypothesis that the Cl^- conductance responsible for salt and fluid secretion in exocrine secretory cells may originate, in part, with the exocytotic fusion of the granule with the luminal membrane (De Lisle and Hopfer 1986). The Cl^- conductance, studied in isolated granules or in permeabilized cells (with intact granules) from rat pancreas and parotid glands, has been described as DIDS-sensitive and showing an anion selectivity sequence of $I^- > Br^- > Cl^- > F^- > SO_4^{2-} >> $ gluconate (De Lisle and Hopfer 1986; Fuller et al. 1989b; Gasser and Hopfer 1990). In addition, the granular membrane also contains a Cl^-/HCO_3^- exchanger (Gasser and Hopfer 1990) and a K^+ conductance which is not sensitive to TEA^+, quinidine and Ba^{2+} in the parotid granules (Gasser et al. 1988a,b). In pancreatic zymogen granules, however, the K^+ conductance is sensitive to Ba^{2+} and quinidine (Gasser et al. 1988b). Thus, a flux of Cl^-, cations and water into the granule would lead to solubilization, flushing out of enzymes and fluid secretion at the time the granular membrane merges with the luminal membrane. The granular fusion and retrieval lead to increases and decreases in the surface area of the luminal membrane, which is reflected by changes in the cell membrane capacitance (Maruyama 1986, 1989). Gasser and Hopfer (1990) postulated that the transport pathways of the granule membrane and the basolateral membrane are sufficient to explain the various types of salivary secretion, i.e., transepithelial NaCl, $NaHCO_3$ and KCl secretion (Fig. 4). Since secretagogues like CCK and secretin stimulate Cl^- transport in zymogen granules isolated from a pre-stimulated pancreas (Gasser et al. 1988a; Piiper et al. 1991), it is reasonable to postulate that regulation of these granular transports would occur via receptor-mediated pathways. The intracellular regulation may involve a Ca^{2+}-mediated pathway, protein kinases A and C and modulation by adenosine and guanidine nucleotides (Gasser et al. 1988a; Fuller et al. 1989a,b; Maruyama 1989; Thévenod et al. 1990). From studies on intact glands, it seems that different agonists, like acetylcholine and isoproterenol, stimulate a release of similar proteins, supporting the view that different regulatory pathways converge on the same secretory granules (Case et al. 1988). Recently, Piiper et al. (1991) found that in the rat pancreas the granular membrane contains G proteins similar to those in the plasma membrane. Additionally, intracellular-intragranular pH gradients and the membrane fluidity may play a modulating role on Cl^- transport activity (Arvan and Castle 1986; Gasser et al. 1988b, 1990; Gasser and Hopfer 1990).

Although the fluid and electrolyte transport models encompassing secretory granules are rather attractive, one important consequence would be that protein and fluid secretion (originating in the acini) are more or less coupled. This might be the case in some glands, e.g., parotid glands secrete large volumes of fluid and enzymes following cholinergic stimulation. In other glands, however, the fluid and enzyme secretions are relatively independent. For example, in the rabbit mandibular gland fluid secretion evoked by cholinergic stimulation is ten times higher than that produced by β-adrenergic stimulation, yet the protein concentration of cholinergic saliva is about ten times lower (Case et al. 1980b; Novak et al. 1984). Even with cholinergic stimulation, the dose of acetylcholine required for optimal fluid secretion is different to that for optimal protein secretion (Case et al. 1980b, 1988). Such a dissociation between fluid and protein secretion is perhaps even greater when one compares salivary glands and the pancreas. Salivary glands, maximally stimulated with acetylcholine, secrete more than 100 µl/g-min, while the pancreas, stimulated with CCK or acetylcholine, secrete maximally about 10 µl/g-min (see Sect. 2.1). The protein secretion is in the order of about 0.5 g/l in salivary glands and about 50 g/l in the pancreas (Case et al. 1980b; Novak et al. 1984; Seow et al. 1986, 1991).

Taken together, although there is no doubt that protein secretion can be accompanied by some fluid secretion (in secretory endpieces), the extent of coupling depends on the gland and the species. However, there must be a protein-independent component of water and electrolyte secretion (in acini or perhaps more distally in ducts). Nevertheless, the idea of the subcellular localization of channels, especially Cl^- channels, remains attractive. Currently, many laboratories are searching for Cl^- channels in epithelia as diverse as airway epithelia, kidney, pancreatic ducts and colonic tumour cell lines which are presumably located in as yet undefined vesicles that may have nothing to do with secretory granules (Decker et al. 1991; Garber 1991; Jilling et al. 1991).

3 Salivary Ducts

Perhaps the greatest differences among salivary glands lie in their duct morphology and composition of the final saliva, which according to Thaysen's two-stage hypothesis is due to the ductal modification of electrolyte components of primary saliva with little water secretion or absorption (see Sect. 1.1). The specific function of various intralobular ducts (intercalated, striated, granular and intralobular excretory ducts) in the production of final saliva remains unclear, no doubt due to their relative inaccessibility for physiological studies. Hence, up to date the gross role of salivary ducts has been derived from two types of studies, those of whole glands and main extralobular ducts.

In whole gland studies, assuming that the primary saliva remains unaltered during stimulation, changes in the ion concentration of final saliva with the secretory rate, so-called "excretory patterns" (Fig. 5), usually reflect the ductal modification. These excretory curves for various glands and their interpretations are well described in several reviews (Young 1979; Young and Schneyer 1981; Young et al. 1987). In general, it can be summarized that in most glands the salivary Na^+ and Cl^- concentrations are low at low secretory rates and increase to approach an asymptote around plasma-like concentrations, an observation which suggests that the ducts reabsorb Na^+ and Cl^- (Fig. 5). In some glands, however, the salivary Na^+ value is high at all secretory rates, most likely due to low ductal transport (e.g., in the cat sublingual gland) (Kaladelfos and Young 1973). In other glands like the sheep parotid, kangaroo parotid and koala mandibular glands, ductal transport is dependent on the Na^+ balance and/or adrenal status of the animal (Compton et al. 1980; Beal 1986, 1991). Usually, the Cl^- concentrations parallel those of Na^+ (Fig. 5), implying that Cl^- transport follows Na^+ transport passively. Exceptions, however, are again the sheep parotid and kangaroo parotid glands, where Cl^- might be secreted in exchange for HCO_3^- (Compton et al. 1980; Beal 1984). The salivary K^+ concentrations are high at low secretory rates and approach and asymptote at or above plasma-like K^+ levels at high secretory rates, suggesting that the ducts secrete K^+. The excretory patterns for HCO_3^- are perhaps the most variable, and it seems that in some glands, ducts secrete this anion, while others absorb it. Some differences in the HCO_3^- excretory curves might arise as a consequence of autonomic stimulation (Young and Martin 1971; Case et al. 1980b; Novak and Young 1986). Furthermore, it appears that in the kangaroo and perhaps the sheep parotid gland some HCO_3^- and fluid secretion originates in the acini (Blair-West et al. 1980; Beal 1990a, 1991b, 1992; see Sect. 2.3.1).

Ductal transport mechanisms, which support most of the above mentioned observations, have been most extensively studied in the main excretory ducts of the rat and rabbit mandibular glands perfused in vivo and in vitro (Young et al. 1987; Cook and Young 1989b; Fig. 6). Although there are some differences in the transport properties and also morphological differences between the two types of ducts, they will be described together. Both epithelia have a low hydraulic conductance (Young and van Lennep 1979; van Os et al. 1981). The luminal membrane contains a large Na^+ conductance sensitive to amiloride and its derivatives, and it discriminates against K^+ and choline but not Li^+ (Knauf and Frömter 1971; Field and Young 1973; Bijman et al. 1981; Young et al. 1987). The principal pathway for Na^+ exit across the basolateral membrane is the $(Na^+ + K^+)$-ATPase. There is good evidence that in the rat duct Na^+ can also enter the cell via the luminal Na^+/H^+ antiport, while in the rabbit ducts the evidence is somewhat contradictory (Knauf and Lübcke 1975; Bijman et al. 1980; Knauf et al. 1982; Young et al. 1987). In the rabbit duct there is clear evidence that the luminal, as well as the basolateral, membrane has a significant Cl^- conductance (Frömter et al. 1974; Augustus et al. 1977, 1978). This transcellular route for Cl^- transport is in many ways similar to that found in sweat ducts (Bijman and Frömter 1986). Both types of epithelia are characterized by a low transepithelial resistance and tight "tight" junctions. Rat ducts are not as permeable to Cl^-, but it is assumed that anion conductances are also located in cell membranes, although no direct studies are available. In sheep and kangaroo parotid glands the Na^+ and K^+ transport depends on the Na^+ body status, i.e., in the Na^+-depleted state, Na^+ absorption and K^+ secretion are increased. It seems, however, that the ducts in these species reabsorb Cl^- in exchange for HCO_3^- independently of the cation transport (Compton et al. 1980; Beal 1984), and hence it has been postulated that the luminal membrane contains a Cl^-/HCO_3^- antiport in series with a basolateral Na^+/H^+ antiport (Young et al. 1987). In the rabbit main duct the sensitivity of the luminal transport to disulphonic stilbenes and HCO_3^- concentration suggests that this membrane also contains a Cl^-/HCO_3^- antiport (Bijman et al. 1980). Both rat and rabbit ducts secrete K^+ against an electrochemical gradient, but the rat ducts are better in this respect. The normal path for K^+ entry across the basolateral membrane is the $(Na^+ + K^+)$-ATPase, and there is a significant K^+ back-leak via the basolateral K^+ conductance (Knauf and Frömter 1971). The luminal membrane is thought to have a low K^+ conductance, but K^+ enters the lumen via a K^+/H^+ antiport, a transport process which was unmasked in the rat duct after the Na^+ transport was prevented (Knauf and Lübcke 1975; Knauf et al. 1975, 1982). The higher capacity of rat ducts to secrete K^+ and HCO_3^- compared with rabbit ducts is ascribed to a greater density of the K^+/H^+ antiports over the Na^+/H^+ antiports. It is also postulated that in a way similar to the cortical collecting ducts, variations in the body acid-base status would be reflected in the ductal HCO_3^- and K^+ secretion and distribution of dark cells in rat extralobular ducts (Knauf et al. 1983).

Apart from the influence of mineralocorticoids and the salt and acid-base statue, the salivary ducts are under autonomic nervous control. There is an

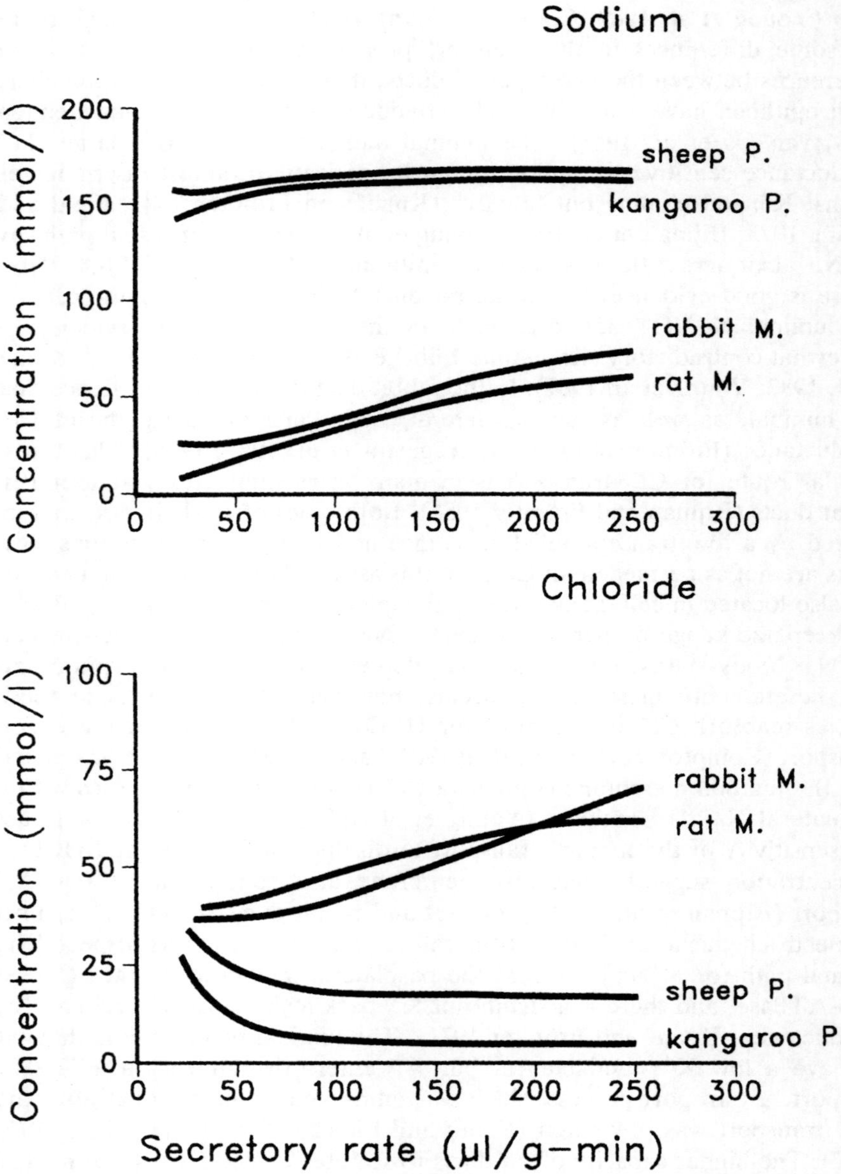

Fig. 5. Electrolyte excretory patterns of cholinergically stimulated salivary glands. Salivary electrolyte concentrations are plotted as a function of salivary secretory rates corrected for gland weight. Data for rat and rabbit mandibular glands are redrawn from Novak (1984), for Na^+-replete sheep parotid glands from Compton et al. (1980) and for Na^+-replete kangaroo parotid glands from Beal (1984)

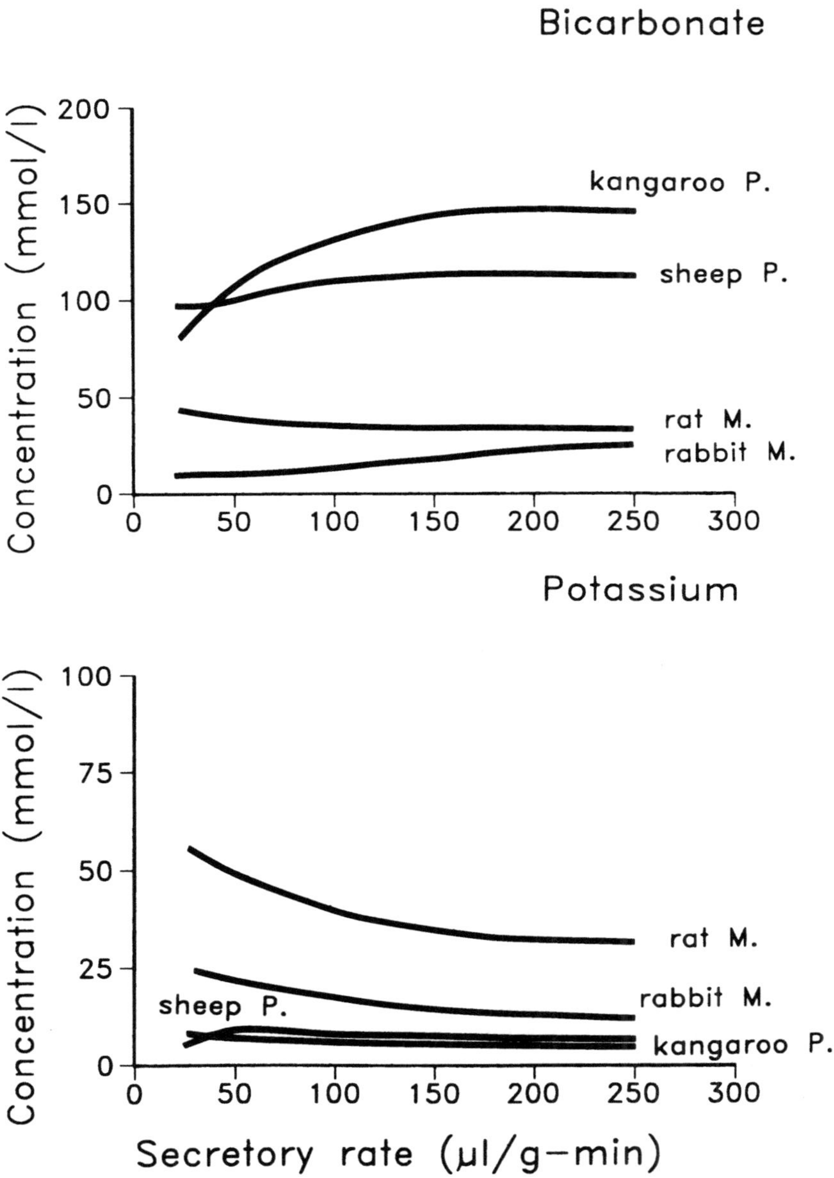

Fig. 5. *Continued*

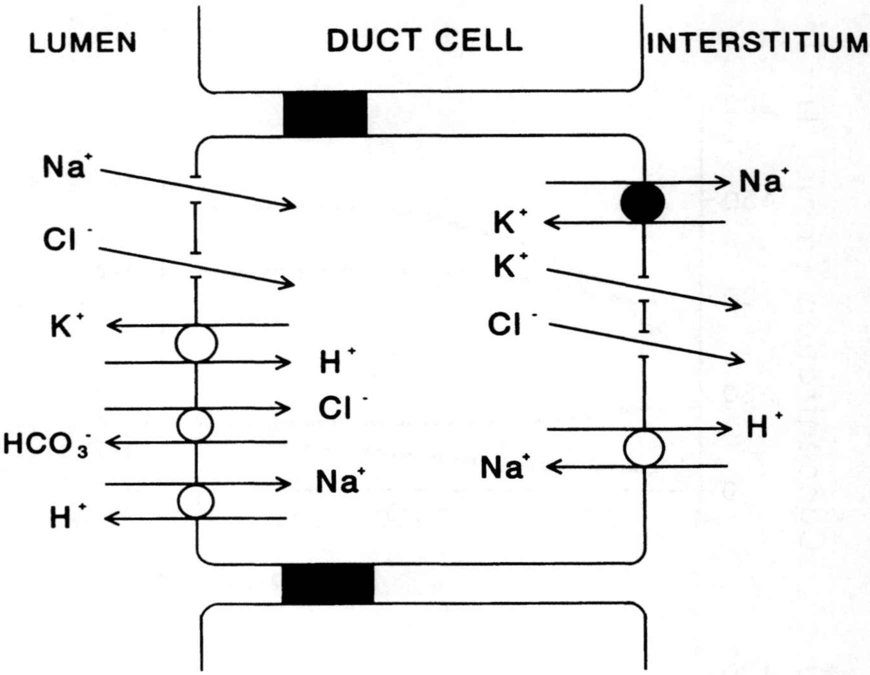

Fig. 6. Model for electrolyte transport based on studies of the main excretory duct of rat and rabbit mandibular glands. This model can account for the Na^+ and Cl^- absorption and K^+ and HCO_3^- secretion seen in some glands, e.g., rat and rabbit mandibular glands (Fig. 5). In order to explain the cation-independent exchange between Cl^- and HCO_3^-, as would appear to be the case for the sheep and kangaroo parotid glands (Fig. 5), luminal Cl^-/HCO_3^- and basolateral Na^+/H^+ antiports have been postulated. As described in Sect. 2.3.1, however, HCO_3^- secretion might take place in the secretory endpieces. The validity of this transport model for intralobular ducts is still to be verified

abundance of morphological data showing that they are innervated by both parasympathetic and sympathetic fibres, and also by fibres containing gastrointestinal peptides, tachykinins and purines (Hökfelt et al. 1977; Young and van Lennep 1978; Uddman et al. 1980; Lundberg 1981; Gallacher 1982). In microperfusion experiments on rat and rabbit main ducts, cholinergic agonists and stimulation of the parasympathetic secretomotor nerves inhibit Na^+ reabsorption (Young et al. 1970; Martin and Young 1971b; Schneyer 1977; Denniss and Young 1978). In rat ducts the K^+ secretion is also inhibited, while HCO_3^- secretion is stimulated. If, however, Na^+ is removed, the secretion of both K^+ and HCO_3^- is stimulated (Young et al. 1970; Martin and Young 1971b). Strong sympathetic nerve stimulation and α-adrenergic drugs have a similar effect to cholinergic stimulation. β-agonists, on the other hand, stimulate K^+ and HCO_3^- secretion (Young and Martin 1971) and a slight Na^+ reabsorption (Martin and Young 1971b). In contrast to studies on main excretory ducts, studies of excretory patterns in perfused rabbit mandibular glands show that an increasing dose of cholinergic agonists increases the ductal Na^+ and Cl^- reabsorption and K^+ secretion (Case et al. 1980b, 1988). Regarding adrenergic stimulation, the rabbit mandibular gland is relatively insensitive; α-adrenergic stimulation evokes no secretory response, and β-adrenergic stimulation evokes only about 12% of the cholinergic response (Case et al. 1980b; Novak 1984). Hence, Case and coworkers (1988) used forskolin and isoproterenol on a background of acetylcholine to study the excretory curves. Forskolin, presumably by increasing cAMP, inhibited Na^+ and Cl^- absorption and K^+ secretion. Isoproterenol acted similarly to forskolin at low doses, but at higher concentrations it increased Na^+ transport. In contrast to the rabbit, the rat mandibular gland secretes well in response to a β-adrenergic stimulation (Fig. 1). Ductal K^+ secretion is markedly enhanced by phenylephrine (to 80–90 mmol/l) and even more so by isoproterenol (to 90–120 mmol/l) (Novak 1984). The effects on Na^+ transport, however, are minimal. Thus, it seems that there are discrepancies between whole gland and main duct studies, which become especially obvious with cholinergic stimulation. This might be explained if in the whole gland the infused agonist causes secondary effects due to the release of biologically active substances present in the gland. Another explanation is that the properties of the main duct do not reflect those of the minor ducts.

Intralobular ducts form the bulk of the ductal epithelial mass and must be responsible for most of the ductal transport processes. However, due to their inaccessibility and the lack of techniques to isolate them, they have not been studied until very recently. Denhaye and Turner (1991) succeeded in isolating granular ducts from rat mandibular glands using isopycric centrifugation following gland digestion. Carbachol and adrenalin (α) stimulation increased the intracellular Ca^{2+} level, and isoproterenol and forskolin increased that of cAMP (Dehaye and Turner 1991; Valdez and Turner 1991). In agreement with their latter finding is the fact that a cAMP-dependent protein kinase was demonstrated by immunocytochemistry in rat parotid intralobular ducts (Schwoch 1987). Substance P, on the other hand, had no effect on Ca^{2+} levels in rat granular ducts, suggesting a lack of receptors for this substance (Valdez and Turner 1991). This finding was rather unexpected, since substance P is found in nerves supplying the salivary glands. Substance P elicits secretion and membrane potential changes in secretory endpieces, and it modifies the electrolyte transport in rat and rabbit main excretory ducts (Hökfelt et al. 1977; Denniss and Young 1978; Coroneo et al. 1979; Martinez and Martinez 1981; Soltoff et al. 1989; Gallacher et al. 1990). The physiological response of intralobular ducts to various agonists and their intracellular messengers has been assessed from their ability to release kallikrein. Thus, carbachol, adrenaline, isoproterenol, but not substance P elicit the release of kallikrein from granular ducts (Dehaye and Turner 1991).

Regarding ion transport processes in intralobular ducts, there are no classical electrophysiological studies as there are in various segments of the nephron and pancreatic ducts (Greger 1981; Novak and Greger 1988a,b) or the main excretory salivary ducts. There are now, however, preliminary reports of patch-clamp studies. In the basolateral membrane of freshly isolated granular ducts from the mouse mandibular gland, two types of K^+ channels have been detected (Dinudom et al. 1991). The first channel observed in cell-attached patches is an inward rectifier activated by membrane depolarization and has a conductance of 140 pS at negative and 24 pS at positive voltages. In the cell-excised mode the channel has a conductance of 240 pS (in symmetrical KCl solution), and the current-voltage relationship becomes linear. Unlike the maxi K^+ channel, this channel is insensitive to Ba^{2+} and TEA^+ but is sensitive to quinidine. The second type of K^+ channel has a characteristic flicker transition between open and closed states, showing rectification and a maximum conductance of 90 pS. In a small percentage of cell-attached patches, a 40 pS anion channel, probably a Cl^- channel, was detected. Non-selective cation channels were also detected, but only in excised patches. Isolated small ducts do not render easy access to the luminal membrane. Some properties of the luminal membrane, however, have been characterized in a HSG cell line, which is derived from intercalated ducts of a human submandibular gland (Ishikawa et al. 1991). An outwardly rectifying Cl^- channel (43 pS at positive voltages) was detected in inside-out patches, but not cell-attached recordings. Its anion permeability sequence and a lack of effect of disulphonic stilbenes suggest that this is not the classical intermediate conductance outwardly rectifying Cl^- channel detected in other epithelia (Tilmann et al. 1991). In inside-out patches of apical membrane (KCl in both pipette and bath), a small 25 pS K^+ channel sensitive to changes in cytosolic Ca^{2+} was detected.

Although patch-clamp studies on small intralobular ducts describe channels which might correlate with the model of electrolyte transport derived from the studies of main extralobular ducts, the physiological relevance of these channels remains to be assessed. These studies must be supported by, for example, classical electrophysiological studies and flux studies, in which the macroscopic conductances, ion permeability and sensitivity to transport inhibitors are characterized.

In order to proceed further in our understanding of transport processes in the salivary glands, it is important to distinguish any functional difference between various morphologically distinct parts of the ductal system. For example, the granular ducts secrete a variety of substances (see Sect. 1.2); it would be surprising if they prove to be purely absorptive and without the capacity to secrete some fluid along with granular secretion. Furthermore, intercalated ducts have been regarded as inert as most micropuncture samples of the primary fluid come from these ducts, since the endpiece lumen is too narrow. Yet these ducts receive a rich supply of close-contact nerves and are, like endpieces, surrounded by contractile myoepithelial cells and are embryologically close to acini. Consequently, defining the role of these ducts, be it secretory, absorptive or neither, is also crucial to our understanding of the secretory processes in salivary glands.

Acknowledgements

I would like to than A.M. Beal for providing some of his unpublished data, C.P. Hansen and E. Schlatter for criticism of the manuscript, and I. Burhoff for preparation of the figures.

References

Adolph EF (1947) Tolerance to heat and dehydration in several species of animals. Am J Physiol 151:564–575

Aloe L, Alleva E, Böhm A, Levi-Montalcini R (1986) Aggressive behaviour induces release of nerve growth factor from mouse salivary gland into the bloodstream. Proc Natl Acad Sci USA 83:6184–6187

Aronson PS (1981) Identifying secondary active solute transport in epithelia. Am J Physiol 240:F1–F11

Arvan P, Castle JD (1986) Isolated secretion granules from parotid glands of chronically stimulated rats possess an alkaline internal pH and inward-directed H^+ pump activity. J Cell Biol 103:1257–1267

Augustus J, Bijman J, van Os CH, Slegers JFG (1977) High conductance in an epithelial membrane not due to extracellular shunting. Nature 268:657–658

Augustus J, Bijman J, van Os CH (1978) Electrical resistance of rabbit submaxillary main duct: a tight epithelium with leaky cell membranes. J Membr Biol 43:203–226

Babkin BP (1944) Secretory mechanism of the digestive glands. Hoeber, New York

Beal AM (1984) Electrolyte composition of parotid saliva from sodium-replete red kangaroos (*Macropus rufus*). J Exp Biol 111:225–237

Beal AM (1986) Responsiveness of the parotid salivary gland of red kangaroos (*Macropus rufus*) to mineralocorticoids. J Exp Biol 120:309–324

Beal AM (1990a) Fluid and bicarbonate secretion by the parotid gland of the red kangaroo. In: Wong PYD, Young JA (eds) Exocrine secretion II. 9th International Symposium on Exocrine Secretion. United League Graphic & Printing, Hong Kong, pp 17–20

Beal AM (1990b) Secretion rates and composition of parotid saliva in the koala (*Phascolarctos cinereus*). J Zool Lond 221:261–274

Beal AM (1991a) Influence of flow rate and aldosterone administration on mandibular salivary composition in the koala (*Phascolarctos cinereus*). J Zool (Lond) 223:265–279

Beal AM (1991b) The effect of carbonic anhydrase inhibitors on secretion by the parotid and mandibular glands of red kangaroos *Macropus rufus*. J Comp Physiol B 161: 611–619

Bijman J, Frömter E (1986) Direct demonstration of high transepithelial chloride-conductance in normal human sweat duct which is absent in cystic fibrosis. Pflügers Arch 407 (Suppl 2):S123–S127

Bijman J, Slegers JFG, van Os CH (1980) Mechanism for HCO_3^- stimulation of NaCl reabsorption in rabbit submaxillary main duct epithelium. In: Schulz I, Sachs G, Forte JG, Ullrich KJ (eds) Hydrogen ion transport in epithelia. Elsevier, Amsterdam, pp 259–264

Bijman J, Cook DI, van Os CH (1981) Effect of amiloride on electrolyte transport parameters of the main duct of the rabbit mandibular gland. Pflügers Arch 398:96–102

Bing J, Poulsen K (1979) In mice aggressive behaviour provokes vast increase in plasma renin concentration, causing only slight, if any, increase in blood pressure. Acta Physiol Scand 105:64–72
Blair-West JR, Fernley RT, Nelson JF, Wintour EM, Wright RD (1980) The effect of carbonic anhydrase inhibitors on the anionic composition of sheep's parotid saliva. J Physiol (Lond) 199:29–44
Bolender RP (1974) Stereological analysis of the guinea pig pancreas. I. Analytical model and quantitative description of non-stimulated pancreatic exocrine cells. J Cell Biol 61:269–287
Bonting SL, De Pont JJHHM, Jansen JWCM (1980) The role of sodium ions in pancreatic fluid secretion in the rabbit. J Physiol (Lond) 309:533–546
Braun PD, Elliot AC, Lau KR (1989) Indirect evidence for the presence of non-specific anion channels in rabbit mandibular salivary gland acinar cells. J Physiol (Lond) 414:415–431
Brettel C (1869) Parotidsecretion des Schafes im Vergleich zur Nierensecretion. Eckhard's Beitr Anat Physiol 4:89–107
Bro-Rasmussen F, Killmann SA, Thaysen JH (1956) The composition of pancreatic juice as compared to sweat, parotid saliva and tears. Acta Physiol Scand 37:97–113
Bundgaard M, Møller M, Poulsen JH (1977) Localization of sodium pump sites in cat salivary glands. J Physiol (Lond) 273:339–353
Burgen ASV (1956) The secretion of potassium in saliva. J Physiol (Lond) 132:20–39
Case RM, Charlton M, Smith PA, Scratcherd T (1980a) Electrolyte secretory processes in exocrine pancreas and their intracellular control. In: Ribet A, Pradayrol L, Susini C (eds) Biology of normal and cancerous exocrine pancreatic cells. INSERM Symposium No 15. Elsevier Biomedical Press, Amsterdam, pp 41–54
Case RM, Conigrave AD, Novak I, Young JA (1980b) Electrolyte and protein secretion by the perfused rabbit mandibular gland stimulated with acetylcholine or catecholamines. J Physiol (Lond) 300:467–487
Case RM, Conigrave AD, Favaloro EJ, Novak I, Thompson CH, Young JA (1982) The role of buffer anions and protons in secretion by the rabbit mandibular salivary gland. J Physiol (Lond) 322:273–286
Case RM, Hunter M, Novak I, Young JA (1984) The anionic basis of fluid secretion by the rabbit mandibular gland. J Physiol (Lond) 349:619–630
Case RM, Howorth AJ, Padfield PJ (1988) Effects of acetylcholine, isoprenaline and forskolin on electrolyte and protein composition of rabbit mandibular saliva. J Physiol (Lond) 406:411–430
Chaundry AP, Cutler LS, Schmutz JA, Sunderraj M, Pierri LK (1985) Development of the hamster submandibular gland. I. The acinar-intercalated duct complex. J Submicrosc Cytol 17:555
Chipperfield AR (1981) Chloride dependence of frusemide- and phloretin-sensitive passive sodium and potassium fluxes in human red cells. J Physiol (Lond) 312:435–444
Coats DA, Denton DA, Goding JR, Wright RD (1956) Secretion by the parotid gland of the sheep. J Physiol (Lond) 131:13–31
Compton JS, Nelson J, Wright RD, Young JA (1980) A micropuncture investigation of electrolyte transport in the parotid glands of sodium-replete and sodium-deplete sheep. J Physiol (Lond) 309:429–446
Compton JS, Martinez JR, Martinez AM, Young JA (1981) Fluid and electrolyte secretion from the isolated perfused submandibular and sublingual glands of the rat. Arch Oral Biol 26:555–561
Conteas CN, McDonough AA, Kozlowski TR, Hensley CB, Wood RL, Mircheff AK (1986) Mapping subcellular distribution of Na^+-K^+-ATPase in the rat parotid gland. Am J Physiol 250:C430–C441
Cook DI, Young JA (1989a) Effect of K^+ channels in the apical plasma membrane on epithelial secretion based on secondary active Cl^- transport. J Membr Biol 110: 139–146
Cook DI, Young JA (1989b) Fluid and electrolyte secretion by salivary glands. In: Forte JG (ed) Handbook of physiology. The gastrointestinal system. Salivary, pancreatic,

and hepatobiliary secretion. Sect. 6, vol III. American Physiological Society, Bethesda, pp 1–23
Cook DI, Young JA (1999) Cation channels and secretion. In: Young JA, Wong PDY (eds) Epithelial secretion of water and electrolytes. Springer, Berlin Heidelberg New York, pp 15–38
Cook DI, Day ML, Champion MP, Young JA (1988a) Ca^{2+} not cyclic AMP mediates the fluid secretory response to isoproterenol in the rat mandibular salivary gland: whole-cell patch-clamp studies. Pflügers Arch 413:67–76
Cook DI, Gard GB, Champion M, Young JA (1988b) Patch-clamp studies of the electrolyte secretory mechanism of rat mandibular gland cells stimulated with acetylcholine or isoproternol. In: Thorn NA, Treiman M, Petersen OH, Thaysen JH (eds) Molecular mechanism in secretion. 25th Alfred Benzon symposium. Munksgaard, Copenhagen, pp 131–151
Cook DI, Poronik P, Young JA (1990a) Characterization of a 25-pS nonselective cation channel in a cultured secretory epithelial cell line. J Membr Biol 114:37–52
Cook DI, Wegmann EA, Ishikawa T, Young JA (1990b) Cation channels in the parotid and mandibular glands of the sheep. In: Wong PYD, Young JA (eds) Exocrine secretion II. 9th Int Symp on exocrine secretion. United League Graphic & Printing, Hong Kong, pp 35–38
Corcelli A, Turner RJ (1991) Role of phospholipids in the binding of bumetanide to the rabbit parotid Na/K/Cl cotransporter. J Membr Biol 120:125–130
Coroneo MT, Denniss AR, Young JA (1979) The action of physalaemin on electrolyte excretion by the rat mandibular and sublingual salivary glands of the rat. Pflügers Arch 381:223–230
Dardick I, Naiberg J, Leung R, Ramjohn S, Christensen H, Burford-Mason A, Henderson WD, Rippstein P (1990) Ultrastructural study of acinar and intercalated duct organization of submandibular and parotid salivary gland. Lab Invest 63:394–404
Decker ER, Calezo MA, Dubinsky WP (1991) Immunolocalization of a 200 kDa protein associated with plasma membrane vesicles enriched in chloride channels. Pediat Pulmon Suppl 6:252
Dehaye JP, Turner RJ (1991) Isolation and characterization of rat submandibular intralobular ducts. Am J Physiol 261:C490–C496
De Lisle RC, Hopfer U (1986) Electrolyte permeabilities of pancreatic zymogen granules: implications for pancreatic secretion. Am J Physiol 250:G489–G496
Denniss AR, Young JA (1978) Modification of salivary duct electrolyte transport in rat and rabbit by physalaemin, VIP, GIP and other enterohormones. Pflügers Arch 376:73–80
Dinudom A, Cook DI, Young JA (1991) Ion channels in salivary duct epithelium. In: Murakami M, Seo Y, Kuwahara A, Watari H (eds) Ionic basis and energy metabolism of epithelial transport – hot aspects from exocrine secretion. National Institute for Physiological Sciences, Okazaki, pp 187–190
Dissing S, Nauntofte B (1990) Na^+ transport properties of isolated rat parotid acini. Am J Physiol 259:G1044–G1055
Dreux C, Imhoff V, Huleux C, Busson S, Rossignol B (1986) Forskolin, a tool for rat parotid secretion studies: ^{45}Ca efflux is not related to cAMP. Am J Physiol 251: C754–C762
Ekström J (1987) Neuropeptides and secretion. J Dent Res 66:524–530
Emmelin N (1953) On spontaneous secretion of saliva. Acta Physiol Scand 30:34–58
Endre ZH, Young JA (1981) Electrolyte content of primary and final saliva of rabbit mandibular glands studies in vivo. In: Zelles T (ed) Saliva and salivation, Adv Physiol Sci vol 28. Pergamon, Akademiai Kiado, Oxford, pp 29–34
Evans LAR, Pirani D, Cook DI, Young JA (1986) Intraepithelial current flow in rat pancreatic secretory epithelia. Pflügers Arch 407:S107–S111
Exley PM, Fuller CM, Gallacher DV (1986) Potassium uptake in the mouse submandibular glands is dependent on chloride and sodium and abolished by piretanide. J Physiol (Lond) 378:97–108
Field MJ, Young JA (1973) Kinetics of Na transport in the rat submaxillary main duct perfused in vitro. Pflügers Arch 345:207–220

Findlay I, Petersen OH (1985) Acetylcholine stimulates a Ca^{2+}-dependent Cl^- conductance in mouse lacrimal acinar cells. Pflügers Arch 403:328–330
Foskett JK (1990) $[Ca^{2+}]_i$ modulation of Cl^- content controls cell volume in single salivary acinar cells during fluid secretion. Am J Physiol 259:C998–C1004
Frizzell RA, Field M, Schultz SG (1979) Sodium-coupled chloride transport by epithelial tissues. Am J Physiol 236:F1–F8
Frömter E, Gebler B, Schopow K, Pockrandt-Hemstedt H (1974) Cation and anion permeability of rabbit submaxillary main duct. In: Thorn NA, Petersen OH (eds) Secretory mechanisms of exocrine glands. Munksgaard, Copenhagen, pp 496–513
Fuller CM, Deetjen HH, Piiper A, Schulz I (1989a) Secretagogue and second messenger-activated Cl^- permeabilities in isolated pancreatic zymogen granules. Pflügers Arch 415:29–36
Fuller CM, Eckhardt L, Schulz I (1989b) Ionic and osmotic dependence of secretion from permeabilised acini of the rat pancreas. Pflügers Arch 413:385–394
Gallacher DV (1982) Are there purinergic receptors on parotid acinar cells? Nature 296:83–86
Gallacher DV, Morris AP (1986) A patch-clamp study of potassium currents in resting and acetylcholine-stimulated mouse submandibular acinar cells. J Physiol (Lond) 373:379–395
Gallacher DV, Morris AP (1987) The receptor-regulated calcium influx in mouse submandibular acinar cells is sodium dependent: a patch-clamp study. J Physiol (Lond) 384:119–130
Gallacher DV, Maruyama Y, Petersen OH (1984) Patch-clamp study of rubidium and potassium conductances in single cation channels from mammalian exocrine acini. Pflügers Arch 401:361–367
Gallacher DV, Hanley MR, Petersen OH, Roberts ML, Squire-Pollard LG, Yule DI (1990) Substance P and bombesin elevate cytosolic Ca^{2+} by different molecular mechanisms in a rat pancreatic acinar cell line. J Physiol (Lond) 426:193–207
Garber SS (1991) Membrane capacitance changes are correlated with $(cAMP)_i$ and activation of Cl^- conductance in T84 cells. Pediat Pulmon Suppl 6:256
Garrett JR (1975) Changing attitudes on salivary secretion: a short history on spit. Proc R Soc Med 68:553–560
Garrett JR (1982) Adventures with autonomic nerves. Perspectives in salivary glandular innervations. Proc R Soc Lond Biol 17:242–253
Garrett JR, Emmelin N (1979) Activities of salivary myoepithelial cells: a review. Med Biol 57:1–28
Gasser KW, Hopfer U (1990) Chloride transport across the membrane of parotid secretory granules. Am J Physiol 259:C413–C420
Gasser KW, DiDomenico J, Hopfer U (1988a) Secretagogues activate chloride transport pathways in pancreatic zymogen granules. Am J Physiol 254:G93–G99
Gasser KW, DiDomenico J, Hopfer U (1988b) Potassium transport by pancreatic and parotid zymogen granule membranes. Am J Physiol 255:C705–C711
Gasser KW, Goldsmith A, Hopfer U (1990) Regulation of chloride transport in parotid secretory granules by membrane fluidity. Biochemistry 29:7282–7288
Geck P, Pietrzyk C, Burckhardt BC, Pfeiffer B, Heinz E (1980) Electrically silent cotransport of Na, K, Cl in Ehrlich cells. Biochim Biophys Acta 600:432–447
Go VLW, Gardner JD, Brooks FP, Lebenthal E, DiMagno EP, Scheele GA (1986) The exocrine pancreas. Biology, pathobilogy, and diseases. Raven Press, New York
Gray MA, Greenwell JR, Argent BE (1988) Secretin-regulated chloride channel on the apical plasma membrane of pancreatic duct cells. J Membr Biol 105:131–142
Gray MA, Pollard CE, Harris A, Coleman L, Greenwell JR, Argent BE (1990) Anion selectivity and block of the small-conductance chloride channel on pancreatic duct cells. Am J Physiol Cell Physiol 259:C752–C761
Gry PTA (1988) Oscillations of free cytosolic calcium evoked by cholinergic and catecholaminergic agonists in rat parotid acinar cells. J Physiol (Lond) 406:35–53
Greger R (1981) Cation selectivity of the isolated perfused cortical thick ascending limb of Henle's loop of rabbit kidney. Pflügers Arch 390:30–37

Greger R, Schlatter E (1984) Mechanism of NaCl secretion in the rectal gland of spiny dogfish (*Squalus acanthias*). I. Experiments in isolated in vitro perfused rectal gland tubules. Pflügers Arch 402:63–75

Greger R, Velázquez H (1987) The cortical thick ascending limb and early distal convoluted tubule in the urinary concentrating mechanism. Kidney Int 31:590–596

Greger R, Schlatter E, Lang F (1983) Evidence for electroneutral sodium chloride cotransport in the cortical thick ascending limb of Henle's loop of rabbit kidney. Pflügers Arch 369:308–314

Grotmol T, Buanes T, Mathisen Ø, Schistad O, Sejersted OM, Raeder MG (1985) Effects of digitoxin and hypokalaemia on pancreatic $NaHCO_3$ secretion and pancreatic Na,K-ATPase activity. Acta Physiol Scand 124:71–80

Hansen CP, Roch B, Kunzelmann K, Greger R (1991) Inhibition of epithelial chloride channels by epithelial channels. Pflügers Arch 419:134

Hazen-Martin DJ, Simson JAV (1987) Electron microscopic immunostaining of nerve growth factor: secretagogue stimulated submandibular glands. Anat Rec 219:171–179

Heidenhain R (1878) Ueber secretorische und tropische Drüsennerven. Pflügers Arch 17:1–67

Hökfelt T, Johanson O, Kellerht JO, Ljungdahl Å, Nilsson G, Nygård A, Pernow B (1977) Immunocytochemical distribution of substance P. In: von Euler US, Pernow B (eds) Substance P. Raven Press, New York, pp 117–145

Holzgreve H, Martinez JR, Vogel A (1966) Micropuncture and histologic study of submaxillary glands of young rats. Pflügers Arch 290:134–143

Imai Y (1965) Study of the secretion mechanism of the submaxillary gland of the dog. Part 2. Effect of exchanging ions in the perfusate on salivary secretion and secretory potential, with special reference to the ionic distribution in the gland. J Physiol Soc Jpn 27:313–324

Imai Y (1976) Physiology of salivary secretion. Front Oral Physiol 2:184–206

Inoue Y, Kanno T (1982) Secretory effect of vasoactive intestinal polypeptide (VIP), adrenaline and carbachol in isolated lobules of rat parotid glands. Biomed Res 3: 384–389

Ishikawa T, Cook DI, Young JA (1991) Ion channels in a human salivary duct cell line (HSG). In: Murakami M, Seo Y, Kuwahara A, Watari H (eds) Ionic basis and energy metabolism of epithelial transport – hot aspects from exocrine secretion. National Institute for Physiological Sciences, Okazaki, pp 185–186

Iwatsuki N, Nishiyama A (1982) Parotid acinar cells: ionic dependence of isoprenaline-evoked membrane potential changes. Pflügers Arch 393:123–129

Iwatsuki N, Maruyama Y, Matsumoto O, Nishiyama A (1985) Activation of Ca^{2+}-dependent Cl^- and K^+ conductances in rat and mouse parotid acinar cells. Jpn J Physiol 35:933–944

Jilling T, Berta G, Kirk KL (1991) The CF gene product (CFTR) plays a regulatory role in endocytosis and exocytosis. Pediat Pulmon Suppl 6:257

Kaladelfos G, Yound JA (1973) Micropuncture and cannulation study of water and electrolyte excretion in the isotonic-secreting cat sublingual salivary gland. Pflügers Arch 341:143–154

Kaladelfos G, Yound JA (1974) Water and electrolyte excretion in the submaxillary gland studied using micropuncture and duct cannulation techniques. Aust J Exp Biol Med Sci 52:67–79

Katoh K, Nakasato M, Nishiyama A, Sakai M (1983) Activation of potassium transport induced by secretagogues in superfused submaxillary gland segments of rat and mouse. J Physiol (Lond) 341:371–385

Katoh K, Kaneko K, Nishiyama A (1986) Effects of adrenergic neurotransmitter on K transport in superfused segments of rat submaxillary gland. Pflügers Arch 406:1–5

Kay RNB (1960) The rate of flow and composition of various salivary secretions in sheep and calves. J Physiol (Lond) 150:515–537

Knauf H, Frömter E (1971) Studies on the origin of the transepithelial electrical potential difference in salivary duct epithelium. In: Giebisch G (ed) Electrophysiology of epithelial cells. Schattauer, Stuttgart, pp 187–206

Knauf H, Lübcke R (1975) Evidence for Na^+ independent active secretion of K^+ and HCO_3^- by rat salivary duct epithelium. Pflügers Arch 361:55–60
Knauf H, Röttger P, Wais U, Baumann K (1975) On the regulatory handling of Na^+, K^+ and H^+ transport by the rat salivary duct epithelium. Fortschr Zool 23:307–321
Knauf H, Lübcke R, Kreutz W, Sachs G (1982) Interrelationships of ion transport in rat submaxillary duct epithelium. Am J Physiol 242:F132–F139
Knauf H, Lübcke R, Röttger P, Baumannn K, Richter G (1983) Relation of dark cells to the transport of H^+/HCO_3^- and K^+ ions: a microperfusion study in the rat submaxillary duct. Kidney Int 23:350–357
Koefoed-Johnsen V, Ussing HH (1958) The nature of the frog skin potential. Acta Physiol Scand 42:298–308
Koob R, Kraemer D, Trippe G, Aebi U, Drenckhahn D (1990) Association of kidney and parotid Na^+,K^+-ATPase microsomes with actin and analogs of spectrin and ankyrin. Eur J Cell Biol 53:93–100
Kunzelmann K, Gerlach L, Fröbe U, Greger R (1991) Bicarbonate permeability of epithelial chloride channels. Pflügers Arch 417:616–621
Lau KR, Case RM (1988) Evidence for apical chloride channels in rabbit mandibular salivary glands. Pflügers Arch 411:670–675
Lau KR, Elliot AC, Case RM (1989) Acetylcholine-induced intracellular acidosis in rabbit salivary gland acinar cells. Am J Physiol 256:C288–C295
Lau KR, Howorth AJ, Case RM (1990) The effects of bumetanide, amiloride and Ba^{2+} on fluid and electrolyte secretion in rabbit salivary gland. J Physiol (Lond) 425: 407–427
Laugesen LP, Nielsen JOD, Poulsen JH (1976) Partial dissociation between salivary secretion and active potassium transport in the perfused cat submandibular gland. Pflügers Arch 364:167–173
Leaf A (1965) Transepithelial transport and its hormonal control in toad bladders. Ergeb Physiol Biol Chem Exp Pharmakol 56:216–263
Lee SI, Turner RJ (1991) Mechanism of secretagogue-induced HCO_3^- and Cl^- loss from rat parotid acini. Am J Physiol 261:G111–G118
Lingard JM, Young JA (1983) β-Adrenergic control of exocrine secretion by perfused rat pancreas in vitro. Am J Physiol 245:G690–G696
Ludwig C (1861) Lehrbuch der Physiologie des Menschen, vol 2. Winter, Leipzig
Lukacs GL, Nanda A, Rothstein OD, Grinstein S (1991) The chloride channel blocker 5-nitro-2-(3-phenylpropyl-amino) bezoic acid (NPPB) uncouples mitochondria and increases the proton permeability of the plasma membrane in phagocytic cells. FEBS Lett 288:17–20
Lundberg A (1957a) The mechanism of establishment of secretory potentials in sublingual gland cells. Acta Physiol Scand 40:35–58
Lundberg A (1957b) Secretory potentials in the sublingual gland of the cat. Acta Physiol Scand 40:21–34
Lundberg A (1957c) Anionic dependence of secretion and secretory potentials in the perfused sublingual gland. Acta Physiol Scand 40:101–112
Lundberg A (1958) Electrophysiology of salivary glands. Physiol Rev 38:21–40
Lundberg JM (1981) Evidence of coexistence of vasoactive intestinal polypeptide (VIP) and acetylcholine in neurones of cat exocrine glands. Acta Physiol Scand 496:1–57
Lundberg JM, Änggård A, Fahrenkrug J (1982) Complementary role of vasoactive intestinal polypeptide (VIP) and acetylcholine for cat submandibular gland blood and secretion. III. Effects of local infusion. Acta Physiol Scand 114:329–337
Manganel M, Turner RJ (1990) Agonist-induced activation of Na^+/H^+ exchange in rat parotid acinar cells is dependent on calcium but not on protein kinase C. J Biol Chem 265:4284–4289
Manganel M, Turner RJ (1991) Rapid secretagogue-induced activation of Na^+/H^+ exchange in rat parotid acinar cells. Possible interrelationship between volume regulation and stimulus-secretion coupling. J biol Chem 266:10182–10188
Mangos JA, McSherry NR (1969) Micropuncture study of sodium and potassium excretion in rat parotid salivary: role of aldosterone (34311). Proc Soc Exp Biol Med 132:787–801

Mangos JA, Braun G, Hamann KF (1966) Micropuncture study of sodium and potassium excretion in the rat parotid saliva. Pflügers Arch 291:99–103

Mangos JA, McSherry NR, Irwin K, Hong R (1973a) Handling of water and electrolytes by rabbit parotid and submaxillary glands. Am J Physiol 225:450–455

Mangos JA, McSherry NR, Nousia-Arvanitakis S, Irwin K (1973b) Secretion and transductal fluxes of ions in exocrine glands of the mouse. Am J Physiol 225:18–24

Mangos JA, Boyd RL, Loughlin GM, Cockrell A, Fucci R (1981) Secretion of monovalent ions and water in ferret salivary glands. A micropuncture study. J Dent Res 60:733–737

Martin CJ, Young JA (1971a) Electrolyte concentrations in primary and final saliva of the rat sublingual gland studied by micropuncture and catherization techniques. Pflügers Arch 324:344–360

Martin CJ, Young JA (1971b) A microperfusion investigation of the effects of a sympathomimetic and a parasympathomimetic drug on water and electrolyte fluxes in the main duct of the rat submaxillary gland. Pflügers Arch 327:303–323

Martin DK, Cook DI, Young JA (1990) Ca^{2+}–activated Cl^- channels in rat mandibular endpiece cells. In: Wong PYD, Young JA (eds) Exocrine secretion II. 9th international symposium on exocrine secretion. United League Graphic & Printing, Hong Kong, pp 87–90

Martinez JR, Cassity N (1983) Effects of transport inhibitors on secretion by perfused rat submandibular gland. Am J Physiol 245:G711–G716

Martinez JR, Cassity N (1984) Effects of ouabain and furosemide on saliva secretion induced by sympathomimetic agents in isolated, perfused rat submandibular glands. Experientia 40:557–559

Martinez JR, Cassity N (1985a) Cl^- requirement for saliva secretion in the isolated, perfused rat submandibular gland. Am J Physiol 249:G464–G469

Martinez JR, Cassity N (1985b) ^{36}Cl fluxes in dispersed rat submandibular acini: effects of acetylcholine and transport inhibitors. Pflügers Arch 403:50–54

Martinez JR, Martinez AM (1981) Stimulatory and inhibitory effects of substance P on rat submandibular secretion. J Dent Res 60:1031–1038

Martinez JR, Holzgreve H, Frick A (1966) Micropuncture study of submaxillary glands of adult rats. Pflügers Arch 290:124–133

Maruyama Y (1986) Ca^{2+}-induced excess capacitance fluctuation studied by phase-sensitive detection method in exocrine pancreatic acinar cells. Pflügers Arch 407: 561–563

Maruyama Y (1989) Control of exocytosis in single cells. NIPS 4:53–56

Maruyama Y, Gallacher DV, Petersen OH (1983) Voltage and Ca^{2+}-activated K^+ channel in baso-lateral acinar cell membranes of mammalian salivary glands. Nature 302:827–829

Melvin J, Moran A, Turner RJ (1988) The role of HCO_3^- and Na^+/H^+ exchange in the response of rat parotid acinar cells to muscarinic stimulation. J Biol Chem 263:19564–19869

Morris AP, Gallacher DV, Fuller CM, Scott J (1987) Cholinergic receptor-regulation of potassium channels and potassium transport in human submandibular acinar cells. J Dent Res 66:541–546

Murakami M, Novak I, Young JA (1986) Choline evokes fluid secretion by perfused rat mandibular gland without desensitization. Am J Physiol 251:G84–G89

Nakahari T, Murakami M, Sasaki Y, Kataoka T, Imai Y, Shiba Y, Kanno Y (1991) Dose effects of acetylcholine on the cell volume of rat mandibular salivary acini. Jpn J Physiol 41:153–168

Nauntofte B, Dissing S (1988a) Cholinergic-induced electrolyte transport in rat parotid acini. Comp Biochem Physiol C 90:739–746

Nauntofte B, Dissing S (1988b) K^+ transport and membrane potentials in isolated rat parotid acini. Am J Physiol 255:C508–C518

Nauntofte B, Poulsen JH (1986) Effects of Ca^{2+} and furosemide on Cl^- transport on O_2 uptake in rat parotid acini. Am J Physiol 251:C175–C185

Nishiyama A, Petersen OH (1974) Membrane potential and resistance measurement in acinar cells from salivary glands in vitro. J Physiol (Lond) 242:173–188

Novak I (1984) Electrolyte and fluid transport in salivary glands. PhD Thesis Sydney University, Sydney
Novak I (1990) Salivary secretion: studies on intact glands in vivo and in vitro. In: Fleischer S, Fleischer B (eds) Methods of enzymology, vol 192. Biomembranes. Cellular and subcellular transport: epithelial cells. Academic Press, San Diego, pp 3–15
Novak I, Greger R (1988a) Properties of the luminal membrane of isolated perfused rat pancreatic ducts: effect of cyclic AMP and blockers of chloride transport. Pflügers Arch 411:546–553
Novak I, Greger R (1988b) Electrophysiological study of transport systems in isolated perfused pancreatic ducts: properties of the basolateral membrane. Pflügers Arch 411:58–68
Novak I, Young JA (1986) Two independent anion transport systems in rabbit mandibular salivary glands. Pflügers Arch 407:649–656
Novak I, Young JA (1989) Acetate stimulates secretion in the rabbit mandibular gland. Pflügers Arch 414:68–72
Novak I, Conigrave AD, Case RM, Young JA (1980) Secretory processes in the rabbit mandibular gland. I. The role of HCO_3^- ions. In: Schulz I, Sachs G, Forte JG, Ullrich KJ (eds) Hydrogen ion transport in epithelia. Elsevier, Amsterdam, pp 243–250
Novak I, Davé C, Young JA (1984) The anionic basis of secretion by rat and rabbit mandibular glands. In: Case RM, Lingard JM, Young JA (eds) Secretion: mechanism and control. Manchester University Press, Manchester, pp 77–80
Okada M, Saito Y, Sawada E, Nishiyama A (1991) Microfluorimetric imaging study of the mechanism of activation of the Na^+/H^+ antiport by muscarinic agonist in rat mandibular acinar cells. Pflügers Arch 419:338–348
Olsen PS, Kirkengaard P, Poulsen S, Nexo E (1986) Vasoactive intestinal polypeptide and acetylcholine stimulate exocrine secretion of epidermal growth factor from the rat submandibular gland. Regul Pept 15:37–46
Osipchuk YV, Wakui M, Yule DI, Gallacher DV, Petersen OH (1990) Cytoplasmic Ca^{2+} oscillations evoked by receptor stimulation, G-protein activation, internal application of inositol trisphosphate or Ca^{2+}: simultaneous microfluorimetry and Ca^{2+} dependent Cl^- current recording in single pancreatic acinar cells. EMBO J 9:697–704
Petersen CCH, Toescu EC, Petersen OH (1991) Different patterns of receptor-activated cytoplasmic Ca^{2+} oscillations in single pancreatic acinar cells: dependence on receptor type, agonist concentration and intracellular Ca^{2+} buffering. EMBO J 10:527–533
Petersen OH (1970a) The dependence of transmembrane salivary secretory potential on the external potassium and sodium concentration. J Physiol (Lond) 210:205–215
Petersen OH (1970b) Some factors influencing stimulation-induced released of potassium from the cat submandibular gland to fluid perfused through the gland. J Physiol (Lond) 208:431–447
Petersen OH (1971) Formation of saliva and potassium transport in the perfused cat submandibular gland. J Physiol (Lond) 216:129–142
Petersen OH (1972) Acetylcholine-induced ion transports involved in the formation of saliva. Acta Physiol Scand 381:1–58
Petersen OH (1980) The electrophysiology of gland cells. Academic Press, London
Petersen OH (1986) Calcium-activated potassium channels and fluid secretion by exocrine glands. Am J Physiol 251:G1–G13
Petersen OH (1987) Electrophysiology of exocrine gland cells. In: Johnson LR (ed) Physiology of the gastrointestinal tract, 2nd edn. Raven Press, New York, pp 745–771
Petersen OH, Gallacher DV (1988) Electrophysiology of pancreatic and salivary acinar cells. Annu Rev Physiol 50:65–80
Petersen OH, Maruyama Y (1984) Calcium-activated potassium channels and their role in secretion. Nature 307:693–696
Petersen OH, Poulsen JH (1967a) The effects of varying the extracellular potassium concentration on the secretory rate and on resting and secretory potentials in the perfused cat submandibular gland. Acta Physiol Scand 70:293–298

Petersen OH, Poulsen JH (1967b) Inhibition of salivary secretion and secretory potentials by g-strophantin, dinitrophenol and cyanide. Acta Physiol Scand 71:194–202

Petersen OH, Poulsen JH (1969) Secretory transmembrane potentials and electrolyte transients in salivary glands. In: Botehlo SY, Brooks FP, Shelly WB (eds) Exocrine glands. University of Pensylvania Press, Philadelphia, pp 3–20

Petersen OH, Wakui M (1990) Oscillating intracellular Ca^{2+} signals evoked by activation of receptors linked to inositol lipid hydrolysis: mechanism of generation. J Membr Biol 118:93–105

Petersen OH, Gallacher DV, Wakui M, Yule DI, Petersen CCH, Toescu EC (1991) Receptor-activated cytoplasmic Ca^{2+} oscillations in pancreatic acinar cells: generation and spreading of Ca^{2+} signals. Cell Calcium 12:135–144

Piiper A, Plusczyk T, Eckhardt L, Schulz I (1991) Effects of cholecystokinin, cholecystokinin JMV-180 and GTP analogs on enzyme secretion from permeabilized acini and chloride conductance in isolated zymogen granules of the rat pancreas. Eur J Biochem 197:391–398

Pirani D, Evans LAR, Cook DI, Young JA (1987) Intracellular pH in the rat mandibular salivary gland: the role of Na-H and Cl-HCO_3 antiports in secretion. Pflügers Arch 408:178–184

Poronnik P, Cook DI, Allen DG, Young JA (1991) Diphenylamine-2-carboxylate (DPC) reduces calcium influx in a mouse mandibular cell line (ST_{885}). Cell Calcium 12: 441–447

Poulsen JH, Bledsoe SW (1978) Salivary gland K^+ transport: in vivo studies with K^+-specific microelectrodes. Am J Physiol 234:E79–E83

Putney JW (1976) Biphasic modulation of potassium release in rat parotid gland by carbachol and phenylephrine. J Pharmacol Exp Ther 198:175–184

Putney JW, Weiss SJ, Leslie BA, Marier SH (1977) Is calcium the final mediator of exocytosis in rat parotid gland? J Pharmacol Exp Ther 203:144–155

Roberts ML, Petersen OH (1978) Membrane potential and resistance changes induced in salivary gland acinar cells by microiontophoretic application of acetylcholine and adrenergic agonists. J Membr Biol 39:297–312

Roberts ML, Iwatsuki N, Petersen OH (1978) Parotid acinar cells: ionic dependence of acetylcholine-evoked membrane potential changes. Pflügers Arch 376:159–167

Royce LS, Baum BJ (1991) Physiologic levels of salivary epidermal growth factor stimulate migration of an oral epithelial cell line. Biochim Biophys Acta 1092:401–403

Schlatter E, Greger R, Weidtke C (1983) Effects of "high ceiling" diuretics on active salt transport in the cortical thick ascending limb of Henle's loop of rabbit kidney. Correlation of chemical structure and inhibitory potency. Pflügers Arch 396:210–271

Schneyer LH (1977) Parasympathetic control of Na, K transport in perfused submaxillary duct of the rat. Am J Physiol 233:F22–F28

Schwoch G (1987) Immunocytochemical demonstration of cyclic AMP-dependent protein kinases in duct cells of the rat parotid. Cell Biol Int Rep 11:599–604

Seow KFT, Lingard JM, Young JA (1986) Anionic basis of fluid secretion by rat pancreatic acini in vitro. Am J Physiol 250:G140–G148

Seow KFP, Case RM, Young JA (1991) Pancreatic secretion by the anaesthetized rabbit in response to secretin, cholecystokinin, and carbachol. Pancreas 6:385–391

Shigetomi T, Hayashi T, Ueda M, Kaneda T, Tokuno H, Takai A, Tomita T (1991) Effects of Ca^{2+} removal and of tetraethylammonium on membrane currents induced by carbachol in isolated cells from the rat parotid gland. Pflügers Arch 419:332–337

Silva P, Stoff JS, Field M, Fine L, Forrest JN, Epstein FH (1977) Mechanism of active chloride secretion by shark rectal gland: role of Na–K–ATPase in chloride transport. Am J Physiol 233:F298–F306

Smaje LH, Poulsen JH, Ussing HH (1986) Evidence from O_2 uptake measurements for Na^+–K^+–$2Cl^-$ co-transport in the rabbit submandibular gland. Pflügers Arch 406: 492–496

Soltoff SP, McMillian MK, Cantley LC, Cragoe EJ, Talamo BR (1989) Effects of muscarinic, alpha-adrenergic, and substance P agonists and ionomycin on ion transport mechanisms in the rat parotid acinar cells. J Gen Physiol 93:285–319

Soltoff SP, McMillian MK, Lechleiter JD, Cantley LC, Talamo BR (1990) Elevation of $[Ca^{2+}]_i$ and the activation of ion channels and fluxes by extracellular ATP and phospholipase C-linked agonists in rat parotid acinar cells. Ann NY Acad Sci 603: 76–92

Stensen N (1662) Nicolai Stenosis Observationes Anatomicae, Quibus Varia Oris, Oculorum, & narium Vasa describuntur, Novique salivae, lacrymarum & muci fintes deteguntur. Lugol Batav Jacobum Chouët, Leiden

Steward MC, Seo Y, Case RM (1989) Intracellular pH during secretion in the perfused rabbit mandibular salivary gland measured by ^{31}P NMR spectroscopy. Pflügers Arch 414:200–207

Stewart DJ, Pon DJ, Sen AK (1983) Cholinergic stimulation of ouabain-sensitive respiration in rat submandibular gland. Am J Physiol 245:G364–G368

Tabcharani JA, Jensen TJ, Riordan JR, Hanrahan JW (1989) Bicarbonate permeability of the outward rectifying anion channel. J Membr Biol 112:109–122

Tabcharani JA, Low W, Elie D, Hanrahan JW (1990) Low-conductance chloride channel activated by cAMP in the epithelial cell line T_{84}. FEBS Lett 270:157–164

Thaysen JH (1960) Handling of alkali metals by exocrine glands other than the kidney. In: Ussing HH, Kruhöffer P, Thorn NA (eds) The alkali metal ions in biology. Handbuch der experimentellen Pharmakologie, vol 13. Springer, Berlin Heidelberg New York, pp 424–463

Thaysen JH, Thorn NA, Schwartz IL (1954) Excretion of sodium, potassium, chloride, and carbon dioxide in human parotid saliva. Am J Physiol 178: 155–159

Thévenod F, Gasser KW, Hopfer U (1990) Dual modulation of chloride conductance by nucleotides in pancreatic and parotid zymogen granules. Biochem J 272:119–126

Tilmann M, Kunzelmann K, Fröbe U, Cabantchik I, Lang HJ, Englert HC, Greger R (1991) Different types of blockers of the intermediate-conductance outwardly rectifying chloride channels in epithelia. Pflügers Arch 418:556–563

Turner RJ, George JN (1988a) Cl^-–HCO_3^- exchange is present with Na^+–K^+–Cl^- contransport in rabbit parotid acinar basolateral membranes. Am J Physiol 254: C391–C396

Turner RJ, George JN (1988b) Ionic dependence of bumetanide binding to the rabbit parotid Na/K/Cl contransporter. J Membr Biol 102:71–77

Turner RJ, George JN (1990) Solubilization and partial purification of the rabbit parotid Na/K/Cl-dependent bumetanide binding site. J Membr Biol 113:203–210

Uddman R, Fahrenkrug J, Malm L, Alumets J, Hakanson R, Rundler F (1980) Neuronal VIP in salivary glands: distribution and release. Acta Physiol Scand 111:31–38

Ussing HH, Eskesen K (1989) Mechanism of isotonic water transport in glands. Acta Physiol Scand 136:443–454

Valdez IH, Fox PC (1991) Interactions of the salivary and gastrointestinal systems. I. The role of saliva in digestion. Dig Dis 9:125–132

Valdez IH, Turner RJ (1991) Effects of secretagogues on cytosolic Ca^{2+} levels in rat submandibular granular ducts and acini. Am J Physiol 261:G359–G363

van Os CH, Wiedner G, Slegers JFG, Bijman J, Wright EM (1981) Hydraulic and electrical conductivities of rabbit gallbladder and submaxillary main duct epithelium. In: Ussing HH, Bindslev N, Lassen NA, Sten-Knudsen O (eds) Water transport across epithelia. Munksgaard, Copenhagen, pp 178–187

Vayro S, Kemp R, Beechey RB, Shirazi-Beechey S (1991) Preparation and characterization of basolateral plasma-membrane vesicles from sheep parotid glands. Biochem J 297:834–848

Velázquez H, Wright FS (1986) Effect of diuretic drugs on Na, Cl, and K transport by rat renal distal tubule. Am J Physiol 250:F1013–F1023

Wakui M, Nishiyama A (1980a) ACh-evoked complex membrane potential changes in mouse submaxillary gland acini. Pflügers Arch 386:251–259

Wakui M, Nishiyama A (1980b) Ionic dependence of acetylcholine equilibrium potentials of acinar cells in mouse submaxillary gland. Pflügers Arch 386:261–267

Wangemann P, Wittner M, Di Stefano A, Englert HC, Lang HJ, Schlatter E, Greger R (1986) Cl^--channel blockers in the thick ascending limb of the loop of Henle. Structure activity relationship. Pflügers Arch 407:S128–S141
Warnock DG, Yee VJ (1981) Chloride uptake by brush border membrane vesicles isolated from rabbit renal cortex. J Clin Invest 67:103–115
Watson AY, Anderson JK, Siminoshi K, Mole JE, Murphy RA (1985) Cellular and subcellular colocalization of nerve growth factor and epidermal growth factor in mouse submandibular glands. Anat Rec 213:365–375
Wegman EA, Reid AM, Cook DI, Titchen DA, Young JA (1988) Potassium channels in secretory cell of the sheep parotid and mandibular glands. Proc Aust Physiol Pharmacol Soc 19:177P
Wharton T (1659) Adenographia: Sive Glandularum Totius Corporis Descriptio. Ravenateiner, Amsterdam
White JF, Imon MA (1983) A role for basolateral anion exchange in active jejunal absorption of HCO_3^-. Am J Physiol 244:G397–G405
Wong MMY, Foskett JK (1991) Oscillations of cytosolic sodium during calcium oscillations in exocrine acinar cells. Science 254:1014–1016
Wright RD, Blair-West JR, Nelson JF, Tregear GW (1984) Handling of phosphate by a parotid gland (ovine). Am J Physiol 246:F916–F926
Wright RD, Blair-West JR, Nelson JF (1986) Effects of ouabain, amiloride, monenesin, and other agents on ovine parotid secretion. Am J Physiol 250:F503–F510
Wright RD, Blair-West JR, Gibson AP, Shulkes A (1991) Evidence against the involvement of vasoactive intestinal peptide in ovine parotid secretion and blood flow. Clin Exp Pharmacol Physiol 18:469–474
Young JA (1979) Salivary secretion of inorganic electrolytes. In: Crane RK (ed) International review of physiology, gastrointestinal physiology III, vol 19. University Park Press, Baltimore, pp 1–58
Young JA, Martin CJ (1971) The effect of a sympatho- and a parasympathomimetic drug on the electrolyte concentrations of primary and final saliva of the rat submaxillary gland. Pflügers Arch 327:285–302
Young JA, Schneyer CA (1981) Composition of saliva in mammalia. Aust J Exp Biol Med Sci 59:1–53
Young JA, Schögel E (1966) Micropuncture investigation of sodium and potassium excretion in rat submaxillary saliva. Pflügers Arch 291:85–98
Young JA, van Lennep EW (1977) Morphology and physiology of salivary myoepithelial cells. In: Crane RK (ed) Gastrointestinal physiology II. International review of physiology, vol 12. University Park Press, Baltimore, pp 105–125
Young JA, van Lennep EW (1978) The morphology of salivary glands. Academic Press, London
Young JA, van Lennep EW (1979) Transport in salivary and salt glands. In: Giebisch G, Tosteson DC, Ussing HH (eds) Membrane transport in biology, vol IVB. Springer, Berlin Heidelberg New York, pp 563–692
Young JA, Martin CJ, Asz M, Weber FD (1970) A microperfusion investigation of bicarbonate secretion by the rat submaxillary gland. The action of a parasympathomimetic drug on electrolyte transport. Pflügers Arch 319:185–199
Young JA, Case RM, Conigrave AD, Novak I (1980) Transport of bicarbonate and other anions in salivary secretion. Ann NY Acad Sci 341:172–190
Young JA, Case RM, Conigrave AD, Novak I, Thompson CH (1981) Secretory processes in the perfused rabbit mandibular gland. In: Zelles T (ed) Saliva and salivation. Adv Physiol Sci vol 28. Pergamon, Akademiai Kiado, Oxford, Budapest, pp 35–46
Young JA, Cook DI, van Lennep EW, Roberts M (1987) Secretion by the major salivary glands. In: Johnson LR (ed) Physiology of the gastrointestinal tract, 2nd edn. Raven Press, New York, pp 773–815

Chapter 2

Mechanism of Nematocyst Discharge and Its Cellular Control

Michio Hidaka

Contents

1 Introduction

Cnidarians possess unique intracellular organelles, cnidae, which discharge by evaginating their tubular contents following certain appropriate stimuli. Every cnida consists of a capsule, a tubule or shaft, or combination of the two, and intracapsular fluid and is contained in a cell called a cnidocyte. Cnidae are divided into three major categories: nematocysts, spirocysts, and

Department of Biology, University of the Ryukyus, Nishihara, Okinawa 903-01, Japan

Advances in Comparative and Environmental Physiology, Vol. 15

ptychocysts (Mariscal 1984). The cell containing a nematocyst, spirocyst, or ptychocyst is called a nematocyte, spirocyte, or ptychocyte, respectively.

Nematocysts as well as other cnidae discharge by eversion of the tubule. The everted tubules penetrate or entangle a prey animal or a potentially predatory animal. Tubules are continuous with the capsule wall, and the junction between the base of a tubule and the capsule wall is covered by a sealing structure in undischarged nematocysts. Anthozoan nematocysts are sealed by three apical flaps, while hydrozoan and scyphozoan nematocysts are sealed by single trapdoorlike opercula. Anthozoan nematocysts often have elongated capsules, while hydrozoan and scyphozoan nematocysts have more round or oval capsules (Mariscal 1984). Thus, nematocysts can be subdivided into two major groups: anthozoan nematocysts and hydrozoan and scyphozoan nematocysts. They are further classified based on the morphological characteristics of the discharged tubule and/or shaft, and some 30 different types of nematocysts have been recognized (Mariscal 1974).

Nematocysts were first observed in the early 17th century, and the phenomenon of nematocyst discharge has been known for about 150 years (see for review, Tardent 1988). The driving force and trigger mechanism of cnida discharge, however, are still incompletely understood. The mechanism of cnida discharge has been studied mostly using nematocysts. Knowledge about nematocysts and their discharge mechanism has accumulated during this decade. In particular, new findings about the volumetric changes of nematocysts during discharge, the biochemical nature of the capsule wall and the intracapsular fluid, and mechanical properties of the capsule wall have contributed much to our understanding of the mechanism of nematocyst discharge.

2 Hypotheses Regarding the Mechanism of Nematocyst Discharge

A variety of hypotheses regarding the mechanism of nematocyst discharge have been reviewed (Mariscal 1974; Gupta and Hall 1984; Tardent 1988). These hypotheses can be divided into three categories: (1) the contractile hypothesis, (2) the osmotic hypothesis, and (3) the tension hypothesis ("stopper" hypothesis). These hypotheses will be briefly described here.

According to the contractile hypothesis, the contraction of some elements in or around a nematocyst capsule increases the intracapsular pressure, causing discharge of the nematocyst. This hypothesis has generally been dismissed, however, since isolated nematocysts without an apparent contractile structure around them can be induced to discharge.

The classical osmotic hypothesis assumes that the intracapsular fluid has a high osmotic pressure but that no intracapsular pressure exists since the capsule wall of resting nematocysts is impermeable to water. At the time of discharge, the capsule wall becomes permeable to water, and the osmotic influx of water increases the intracapsular pressure, thus leading to discharge of the nematocyst. It has been shown, however, that the capsule wall of resting nematocysts is permeable to dyes whose relative molecular mass is less than 600; hence, the capsule wall is permeable to water and small ions (Lubbock and Amos 1981; Salleo 1984; Weber et al. 1987a). Thus, the hypothesis assuming changes in the permeability of the capsule wall has been dismissed. A modified version of the osmotic hypothesis was proposed by Lubbock and his colleagues; it assumes that the osmotic properties of the intracapsular fluid change at the time of discharge. It has been suggested that the removal of calcium from the nematocysts somehow increases the osmotic pressure of the intracapsular fluid (Lubbock and Amos 1981; Lubbock et al. 1981).

The tension hypothesis assumes that there is a preexisting intracapsular pressure in the nematocysts and that the capsule of resting nematocysts is in an elastically expanded state (Yanagita 1943; Tardent and Holstein 1982; Holstein and Tardent 1984; Salleo 1984). It also assumes that there are springlike intrinsic forces which are built up during cnidogenesis and stored in the tubule (Carré 1980; Godknecht and Tardent 1988). The elastic energy stored in the capsule wall and the tubule provides at least partly the driving force required for nematocyst discharge (Tardent 1988). It has not been demonstrated, however, whether the "stopper", a sealing structure of the tubule, is softened at the time of discharge ("stopper" hypothesis) (Yanagita 1943; Salleo 1984) or whether a temporary increase in the osmotic pressure of the intracapsular fluid triggers the nematocyst discharge (Holstein and Tardent 1984).

3 Role of Calcium in Nematocyst Discharge

Since high concentrations of calcium have been detected in undischarged nematocysts and are lost during the discharge (Lubbock et al. 1981), calcium may have some role to play in the mechanism of nematocyst discharge or in functions of nematocysts. Calcium has been thought to be involved in nematocyst discharge in two ways: A loss of calcium from the capsule increases the osmotic pressure of the intracapsular fluid, or the loss of calcium makes the inverted tubule and the "stopper" region of the capsule unstable. On the other hand, the idea that calcium is not necessarily involved in nematocyst discharge has recently been proposed.

3.1 A Loss of Calcium Increases the Osmolarity of the Intracapsular Fluid

The modified osmotic hypothesis proposed by Lubbock and Amos (1981) and Lubbock et al. (1981) is based on the findings that undischarged holotrichous isorhiza nematocysts of the sea anemones *Rhodactis rhodostoma* and *Anthopleura elegantissima* contain a high concentration of calcium and that it is lost during discharge. Lubbock et al. (1981) and Gupta and Hall (1984) analyzed frozen sections of quenched mesenterial filaments of *R. rhodostoma* and acrorhagi, a specific structure used for aggression, of *A. elegantissima* with an electron microprobe X-ray analyzer. Mature nematocysts contain a fluid rich in calcium (500–600 mmol/kg wet weight). Just before discharge, the nematocyst moves to the surface of the nematocyte and comes into contact with the external sea water (Lubbock et al. 1981). At the beginning of discharge, there is a massive efflux of calcium from the nematocyst fluid that is disproportionate to the loss of dry mass. The nematocyst at this stage of discharge is surrounded by a thin layer of fluid, and the calcium concentration in the fluid exceeds that of the intracapsular fluid of the same nematocyst. The other elemental composition of the surrounding fluid is similar to that of sea water, suggesting that the calcium escapes from the capsule into sea water. As the calcium is lost, there is an influx of sodium and chloride, but not of potassium. This indicates that sea water enters the capsule during the discharge.

The following findings by Lubbock and Amos (1981) also support the idea that calcium is involved in nematocyst discharge: (1) isolated nematocysts of *R. rhodostoma* discharge in a solution of 10 m*M* potassium citrate, but the discharge is inhibited by the addition of 50 m*M* $CaCl_2$ to the citrate solution. Citrate is generally considered to chelate calcium in nematocysts. (2) Isolated nematocysts shrink when immersed in 50 m*M* $CaCl_2$ in 40% w/v polyethylene glycol, but re-expand when placed in a solution which contains 50 m*M* sodium citrate instead of $CaCl_2$. Polyethylene glycol was added to the solutions to prevent nematocyst discharge by increasing the osmolarity of the external solutions. The effect of calcium on the volume of isolated nematocysts was confirmed with basitrichous isorhizas of the sea anemone *Calliactis tricolor* (Fig. 1; Hidaka 1992).

From these observations, Lubbock and Amos (1981), Lubbock et al. (1981), and Gupta and Hall (1984) suggested the following scheme of nematocyst discharge (Fig. 2). (1) In an undischarged nematocyst, polypeptides are bound by calcium ions into polypeptide chains or protein aggregates (Fig. 2A). Upon an appropriate stimulus, the nematocyst moves to the surface of the nematocyte, leading to an exocytotic fusion of the nematocyst membrane with the apical cell membrane. (2) Some unknown agent causes the release of calcium from the capsule fluid. Gupta and Hall (1984) suggested that a change in pH or some other factor shifts the binding constant of calcium to the polypeptide chains in the capsule. The release of

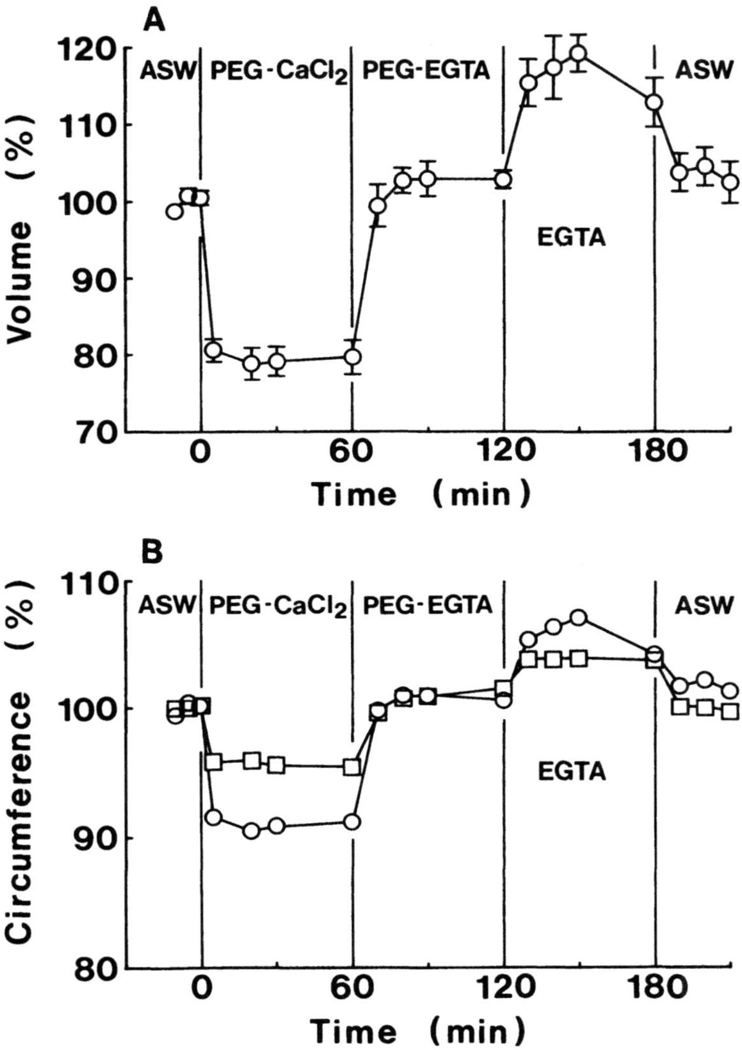

Fig. 1A,B. Effects of Ca^{2+} on the volume (**A**) and circumference (**B**) of nematocysts (basitrichous isorhizas) isolated from the acontium of *Calliactis tricolor*. The nematocysts were immersed in 50 m*M* $CaCl_2$ containing 40% polyethylene glycol (*PEG*), 5 m*M* ethylene glycol-bis(β-aminoethylether)-N,N,N′,N′-tetra-acetic acid (*EGTA*) containing 40% polyethylene glycol, 5 m*M* EGTA, and finally artificial sea water (*ASW*). The volume and circumference are expressed as a percentage of the mean volume and circumference of the nematocysts in ASW. *Vertical bars* in **A** represent standard errors of the means. *Circles and squares* in **B** represent shorter and longer circumferences, respectively. $n = 6$, but 1 of the nematocysts was lost in 5 m*M* EGTA. (Hidaka 1992)

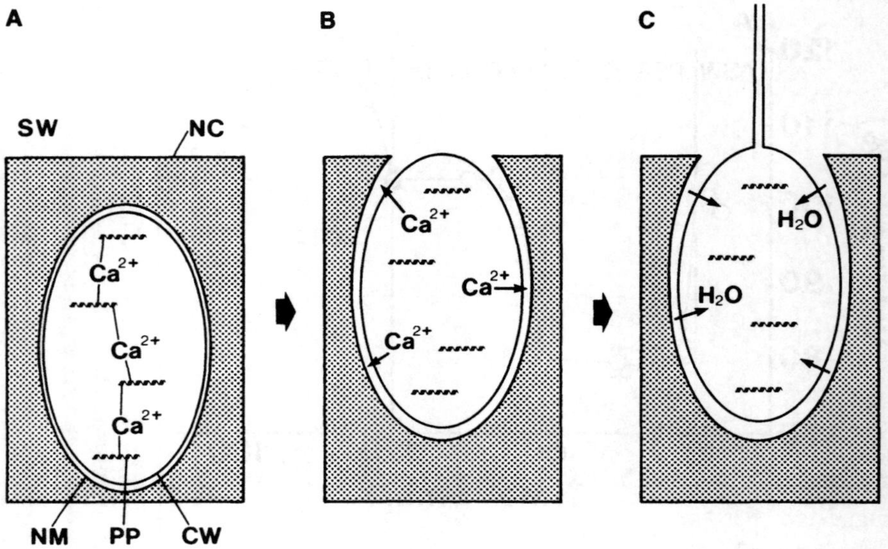

Fig. 2A–C. A schematic diagram of the process of nematocyst discharge based on the modified osmotic hypothesis proposed by Lubbock and his colleagues. **A** Polypeptides (*PP*) are cross-linked by Ca^{2+} to form protein aggregates. **B** A loss of Ca^{2+} dissociates the protein aggregates into low molecular weight polypeptides, thus increasing the internal osmotic pressure. **C** Osmotic influx of water increases the internal pressure leading to the discharge of the nematocyst. *CW* Capsule wall; *NC* nematocyte; *NM* nematocyst membrane; *SW* sea water

calcium from the nematocyst dissociates the polypeptide chains to increase the number of osmotically active molecules (Fig. 2B). (3) Osmotic influx of water (sea water in this case) increases the internal pressure, thus leading to discharge of the nematocyst (Fig. 2C). It has not yet been demonstrated, however, that the release of calcium from the capsule actually increases the osmotic pressure of the intracapsular fluid.

3.2 Calcium Stabilizes the Capsule Wall and the Tubule

Trypsin induces the discharge of isolated nematocysts (Yanagita and Wada 1954; Yanagita 1959a), and this discharge is inhibited by prior exposure to calcium ions (Salleo et al. 1983). Since trypsin cannot permeate the capsule wall because of its high molecular weight, Salleo et al. (1983) proposed that the stabilizing effect of calcium may be exerted on the external surface of the resting nematocyst rather than on the intracapsular fluid. It has, however, not been demonstrated that there are calcium binding sites in the stopper region of the capsule wall. X-ray mapping shows that the calcium is evenly distributed in the capsule of isolated basitrichous isorhizas and is not localized at the capsule wall or in the "stopper" region (Fig. 3). In *Hydra* nematocysts, more than 85% of the divalent cations are found in the soluble extract (Weber et al. 1987b) and are thus not bound to insoluble structural components such as the capsule wall or the tubule.

Salleo et al. (1988a) investigated the time course of the release of Ca^{2+} from isolated nematocysts using a photoprotein, aequorin. They observed two components of light emission when a hypotonic solution containing aequorin was added to a suspension of microbasic mastigophores of *Aiptasia mutabilis*: a rapid release of Ca^{2+} which precedes nematocyst discharge and a slow one which increases in parallel with the discharge rate. The rapid component was lacking when the nematocysts failed to discharge. They suggested that the rapid component of light emission is due to Ca^{2+} released from the sealing structures of the nematocysts and that the release of Ca^{2+} is associated with the discharge. Most nematocysts in their preparation, however, did not discharge immediately after the fast component of light emission but discharged rather gradually. Salleo et al. (1988a) further suggested that the slow light emission is due to Ca^{2+} in the intracapsular fluid ejected during discharge. Slow light emission was also observed in unresponsive nematocysts. This slow component was considered to be due to Ca^{2+} diffusing through the capsule wall. When holotrichous isorhizas

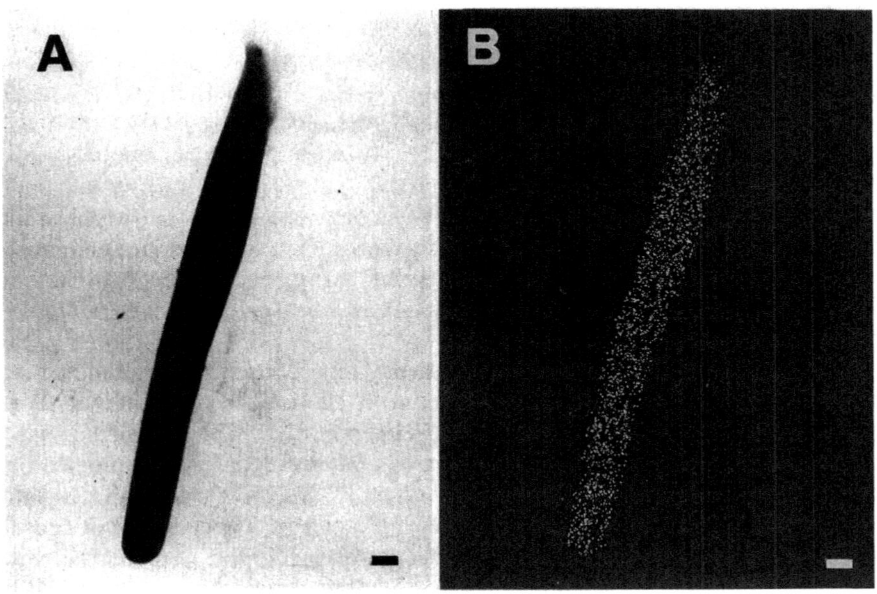

Fig. 3A,B. A scanning transmission electron micrograph of a basitrich nematocyst isolated from the acontium of the sea anemone *Calliactis polypus* in distilled water (**A**) and calcium K X-ray intensity distribution over the same field (**B**). *Scale bar* 1 μm

from the scyphozoan *Pelagia noctiluca* were used, Salleo et al. (1988b) found only the fast component and a very weak slow component. Basitrichous isorhizas isolated from the sea anemone *Calliactis parasitica* do not release measurable amounts of Ca^{2+} (Salleo et al. 1990). If calcium in the capsule is readily replaced by cations in the surrounding medium, as suggested by Weber (1989), care should be taken in the procedure for isolating nematocysts.

Watson and Mariscal (1985) showed that the everted tubule is lined on both sides by subunits that are stained by antimonate. They suggested that these subunits bind calcium and hold the uneverted tubule in its folded and coiled configuration. A dissociation of the calcium from the subunits or rupture of the hydrogen bonds would cause subunit uncoupling and permit the tubule to evert. On the other hand, Gupta and Hall (1984) observed that an uneverted tubule contains zinc and loses most of it during eversion. They suggested that the allosteric constraints on the uneverted tubules might be provided by the salt-bridges mediated by zinc.

3.3 Calcium Is Not Necessarily Involved in Nematocyst Discharge

Isolated, dried nematocysts have been examined using X-ray microanalysis (Mariscal 1980, 1984, 1988; Weber et al. 1987b). Weber et al. (1987b) found that besides sulfur, calcium and magnesium are by far the most abundant atoms in nematocysts isolated from various cnidarians. The ratio between calcium and magnesium varies from one species to another. Weber et al. (1987b) showed by the titration method and atomic absorption measurements that in *Hydra* nematocysts Ca^{2+} (0.36 μmol/mg dry nematocysts) and Mg^{2+} (0.80 μmol/mg dry nematocysts) add up to a total in situ concentration of 0.5–1.0 *M*. Mariscal (1988) also found that in scyphozoan holotrichs and euryteles calcium and magnesium are two dominant elements in addition to sulfur. Figure 4 shows typical X-ray energy spectra obtained from isolated, dried basitrich nematocysts of the sea anemone *Calliactis polypus*.

Calcium and magnesium are considered to be bound to large molecules in the nematocyst since thorough washing of the isolated nematocysts does not result in a loss of these cations (Weber et al. 1987b; Mariscal 1988). However, when isolated nematocysts from *Hydra vulgaris* are immersed in high concentrations (1 *M*) of salts such as KCl, NaCl, and NH_4Cl, the naturally occurring cations are readily replaced by the cations added to the external medium (Weber 1989). In many cases, the nematocysts with newly introduced cations retain their discharge capabilities. Weber (1989) suggested that nematocyst cations become equilibrated immediately with the cations of the surrounding medium and that the isolation procedure and further treatment govern the final ionic composition of isolated nematocysts.

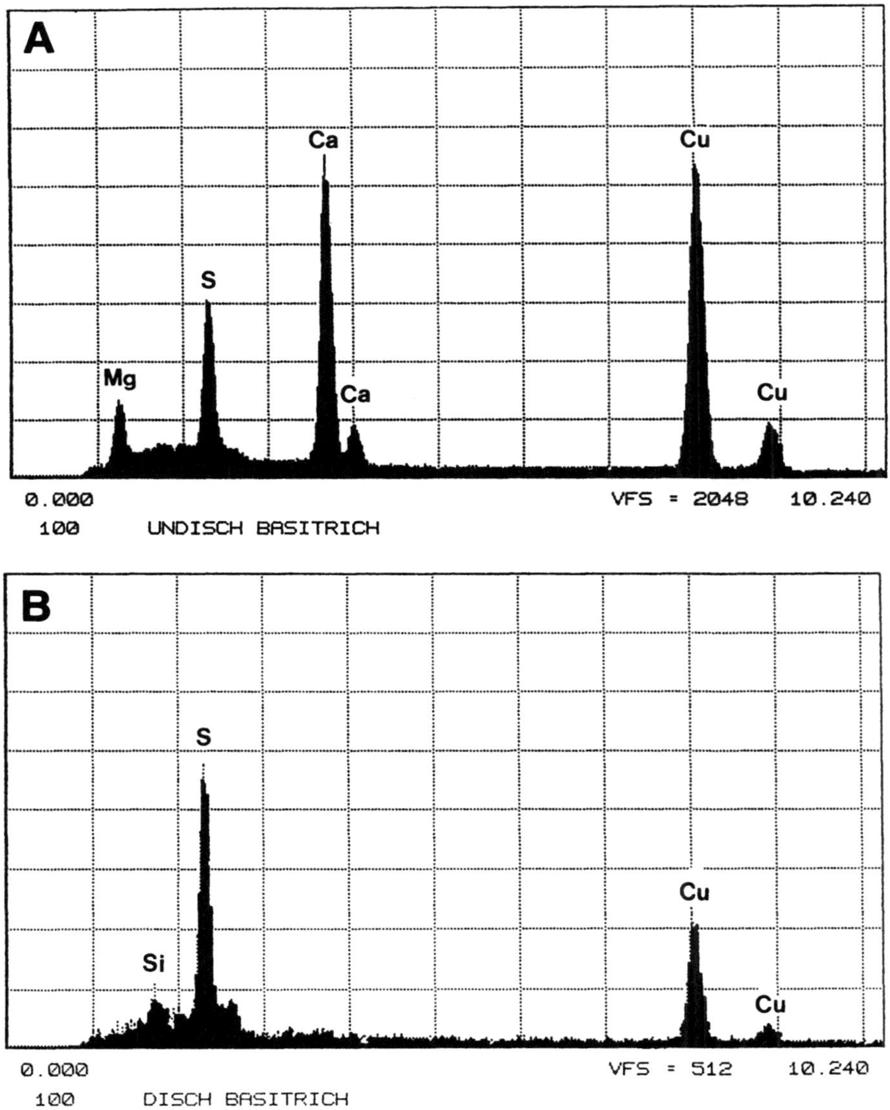

Fig. 4A,B. X-ray energy spectra showing the chemical elements in the undischarged (**A**) and discharged (**B**) basitrich nematocysts isolated from acontia of the sea anemone *Calliactis polypus*. The Cu peaks were derived from the copper mesh

Recent X-ray microanalysis of ultrathin cryosections of *Hydra* nematocysts indeed showed that undischarged and discharging nematocysts contain high concentrations of potassium (more than 2 mol/kg dry weight) instead of calcium and magnesium and that potassium is not lost until the tubule is fully everted (Zierold et al. 1989; Tardent et al. 1990a; Gerke et al. 1991). These observations demonstrate that the naturally occurring cations in *Hydra* nematocysts are not calcium but potassium and suggest that calcium is not necessarily involved in the discharge of, at least, *Hydra* nematocysts.

Mariscal (1984, 1988) found that undischarged holotrichous isorhizas isolated from the feeding tentacles of *Haliplanella luciae* contain large amounts of phosphorus instead of calcium, while the same type of nematocysts isolated from catch tentacles contain calcium. The nonvenomous spirocysts of *H. luciae* feeding tentacles appear to lack both calcium and phosphorus. Furthermore, Mariscal (1984, 1988) and Watson and Mariscal (1985), using X-ray microanalysis and antimonate staining, found that in partially discharged nematocysts the capsule and the lumen of the partially everted tubule contain high concentrations of calcium. They suggest that significant amounts of calcium pass into the everting tubule rather than diffuse outwards through the capsule wall into the surrounding medium as suggested by Lubbock et al. (1981). Mariscal (1988) suggested that calcium is injected into the target tissue with the venom and causes the deleterious effect simply by flooding the target tissue with calcium ions or by stimulating the enzymatic activities of nematocyst venoms. Hemolysin (phospholipase A) of *Aiptasia pallida* requires Ca^{2+} for its activity (Hessinger and Lenhoff 1973), while a hemolytically active fraction of *Hydra* nematocysts does not depend on Ca^{2+} (Klug et al. 1989). It is still to be proved whether calcium is extruded through the tip of the everted tubule.

4 Mechanics of Nematocyst Discharge

4.1 Elastic Energy Stored in Undischarged Nematocysts

The volume of the capsule of nematocysts decreases by 40–50% after discharge (Yanagita 1943; Tardent and Holstein 1982; Holstein and Tardent 1984; Salleo et al. 1986). Tardent and Holstein (1982) examined the volumetric changes of stenoteles of hydra during discharge. They found that the capsule shrinks to 50% of the original volume after discharge and that the total volume of the discharged nematocyst, which includes the volume of the everted tubule and stylets, decreases by 9.3% after discharge. No wrinkle is observed on the capsule wall of shrunken nematocysts, suggesting that the capsule wall of undischarged nematocysts is elastically expanded (Phelan and Blanquet 1985).

Holstein and Tardent (1984) examined the volumetric changes of hydra stenoteles during discharge by means of ultrahigh-speed microcinematography. The nematocyst first everts three stylets, which join to form an arrowhead. Then the stylets withdraw, and a tubule evaginates. The entire discharge process of the stenoteles is completed in about 3 ms. The tip of the stylets travels a distance of approximately 20 μm within 10 μs. This indicates that the stylets are accelerated at about 40 000 *g*. In the same period of 10 μs, the circumference of the capsule shrinks by 11%. The speed of deformation is 11 000 lengths/s and is 500–1000 times faster than that of the fastest striated muscle. This demonstrates that the deformation of the capsule cannot be produced by ordinary contractile elements and strongly suggests that the elastic energy stored in the expanded capsules provides the driving force required for nematocyst discharge.

Robson (1973) observed that holotrichous isorhizas of *Corynactis* mesenterial filaments swell up to 150% on exposure to sea water just before discharge. The capsules return to their original dimensions after discharge. This suggests that the capsule of the nematocysts is not under tension when they stay within the nematocytes, contrary to the stenoteles of *Hydra* tentacles (Holstein and Tardent 1984). The nematocyst capsule begins to expand to store elastic energy only after it comes into contact with sea water. This points out the possibility that isolated nematocysts are different from in situ nematocysts, both in terms of ionic composition and the degree of swelling. Whether the elastic energy is stored in in situ nematocysts or only in those exposed to external solutions, it seems certain that the elastic energy is released during a short period of discharge.

The idea that capsules of isolated nematocysts store a high potential energy is consistent with the observations that isolated nematocysts can be induced to discharge by various agents which may act on the "stopper" of nematocysts. The agents are proteases such as trypsin (Yanagita and Wada 1954; Yanagita 1959a; Salleo et al. 1983) and subtilisin (Weber 1989), disulfide reducing agents such as thioglycolate and dithioerythritol (Yanagita and Wada 1954; Yanagita 1959a; Salleo et al. 1986; Weber et al. 1988), chaotropic anions such as thiocyanate (SCN^-) which weaken hydrophobic interactions (Yanagita 1959b; Salleo et al. 1986; Weber et al. 1988), surface active reagents (Yanagita 1960b), and acids and alkalis whose pH are below 2–3 or above 11 (Yanagita and Wada 1953; Salleo et al. 1984a).

Weber (1989) estimated the elastic moduli of the capsule wall of hydra stenoteles. He measured changes in the volume of nematocysts immersed in solutions whose osmolarity was varied by adding various amounts of polyethylene glycol. The elastic modulus of the capsule wall of isolated stenotele is 0.3–0.4 GPa. Hidaka (1990) also estimated the elastic modulus of the capsule wall of basitrichous isorhizas of *Calliactis polypus* to be of the order of 1 GPa. These values are similar to the elastic modulus (0.8 GPa) of rat tail tendon, which is composed of parallel-running collagen fibers (Rigby et al. 1959). This coincidence is not surprising since the capsule and tubule proteins of nematocysts are included in the collagen class (Blanquet and Lenhoff 1966; Phelan and Blanquet 1985). Once the elastic modulus of the capsule wall is known, the elastic energy stored in the capsule wall can be calculated. The volume of relaxed capsules is estimated to be about 55% of the original volume, that is, the volume of isolated, intact nematocysts. The elastic energy stored in the capsule of 100% volume was calculated to be about 4 J/g hydrated nematocysts (Weber 1989). Weber (1989) suggested that a considerable part of this energy is used to overcome the frictional resistances during the eversion of the tubule.

Tubules in undischarged nematocysts are in a highly twisted state and may store elastic energy which provides, at least partly, the driving force of evagination (Carré 1980; Tardent and Holstein 1982; Godknecht and Tardent 1988). The everted tubule of hydra stenoteles is about 40% longer than the tubule within the undischarged capsule (Tardent and Holstein 1982). During the discharge of microbasic mastigophores of the sea anemone *Anemonia sulcata*, the length and circumference of the tubules increase by 95 and 17%, respectively (Godknecht and Tardent 1988). The elongation of evaginated tubules is due to untwisting rather than stretching of the tubule wall. The increase in the circumference must take place at the tip of the evaginating tubule and reduces the frictional resistances which develop between the outer, evaginated part and the inner, unevaginated part of the tubule during evagination.

4.2 Mechanical Properties of the Capsule Wall and Internal Osmotic Pressure

Swelling of nematocysts just before discharge has been observed both in in situ nematocysts (Robson 1973; Holstein and Tardent 1984) and in isolated nematocysts (Salleo et al. 1986). Holstein and Tardent (1984) reported that when *Hydra* stenoteles were triggered by an electrical pulse the volume of the cysts increased by 10% before the cyst's operculum flipped open. They suggested that the osmolarity of the intracapsular fluid increases just before discharge and that the capsule swells due to the osmotic influx of water. Salleo et al. (1986) showed that when isolated holotrichous isorhizas of *Pelagia noctiluca* are treated with SCN^- the volume of the capsule increases by up to 20% before discharge. They suggested that the initial swelling could be due to the hydration of protein structures of the nematocysts rather than to an osmotically induced water influx. It is also possible that the swelling of nematocysts just before discharge is due to changes in the mechanical properties of the capsule wall.

Nematocysts isolated from acontia of the sea anemone *Calliactis tricolor* usually discharge in 5 m*M* ethylene glycol-bis(β-aminoethylether)-N,N,N′,N′-tetra-acetic acid (EGTA; Hidaka and Mariscal 1988). When the nematocysts remain undischarged in 5 m*M* EGTA, their volume increases up to 115–120% (Fig. 1; Hidaka 1992). The capsule wall of nematocysts is composed mainly of collagenlike protein (Blanquet and Lenhoff 1966). Certain collagenous connective tissues of echinoderms become softened when immersed in Ca-free solutions (Hidaka 1983; Motokawa 1984). Thus, it is not surprising if the mechanical properties of the capsule wall are also affected by Ca^{2+}. To examine whether Ca-free solutions increase the osmotic pressure of the intracapsular fluid or soften the capsule wall, the volumetric changes in isolated nematocysts immersed in solutions containing various concentrations of polyethylene glycol were measured with or without Ca^{2+} (Fig. 5). The nematocysts shrank as the osmolarity of the external solution was raised. The slope of the regression line of the relative volume on the osmolarity of the bathing medium indicates how easily the capsule wall can be expanded, that is, the stiffness of the capsule wall. The slope of the regression line was not significantly affected by Ca^{2+}, suggesting that Ca^{2+} does not affect the elastic modulus of the capsule wall (Hidaka 1990). Weber (1989) estimated the modulus of the capsule wall of hydra stenoteles whose naturally occurring cations were replaced by various other cations. The modulus of stenoteles loaded with Ca^{2+} is about half of that of stenoteles loaded with Na^+ or Mg^{2+} and of untreated stenoteles. Thus, the removal of Ca^{2+} does not soften the capsule wall in hydra stenoteles

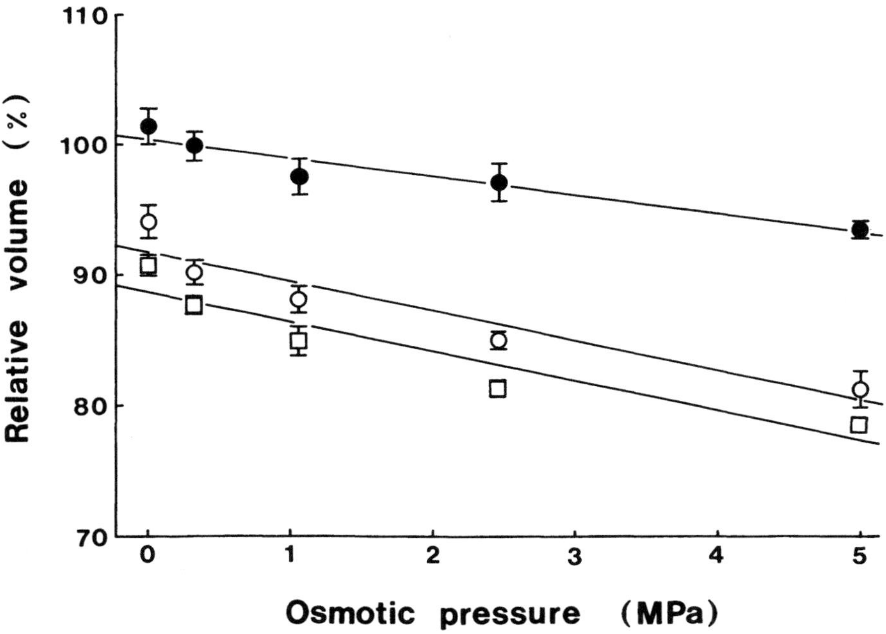

Fig. 5. Effects of Ca^{2+} on the mechanical properties of the capsule wall of basitrich nematocysts isolated from *Calliactis polypus*. Effects of the osmolarity of the external medium on the volume of isolated nematocysts were determined in Ca-free artificial sea water (●), 50 m*M* $CaCl_2$ (○), and 500 m*M* $CaCl_2$ (□). The osmolarity of the external medium was changed by adding 10, 20, 30 and 40% (w/v) polyethylene glycol to the test solutions. Volume of nematocysts was expressed as a percentage of the original volume in ASW. Slope of the regression line of the relative volume on the osmolarity of the external medium was not significantly affected by Ca^{2+}. Means ± SEM

also. These observations suggest that swelling and probably discharge of isolated nematocysts in Ca-free solutions are due to increased osmotic pressure of the intracapsular fluid rather than softening of the capsule wall.

Weber (1989) calculated the internal osmotic pressure of isolated stenoteles. The internal osmotic pressure of untreated stenoteles amounts to 5.5 osmol (13.2 MPa), which is close to the "notional" osmotic pressure of the capsular fluid of holotrichous isorhizas of *Corynactis* determined by the depression of freezing point (Picken and Skaer 1966). The osmolarity differences across the capsule wall at a stenotele volume of 100% were calculated to be 8.1, 5.6, and 2.4 osmol for Na^+-, Mg^{2+}-, and Ca^{2+}-loaded stenoteles, respectively (Weber 1989). Thus, Ca^{2+} is more effective than Na^+ and Mg^{2+} in lowering the internal osmotic pressure. This suggests that Ca^{2+} forms highly coordinated complexes with the intracapsular fluid, thus reducing the internal osmotic pressure.

5 Biochemical Nature of the Capsule and Intracapsular Fluid

The capsule wall and the tubule of nematocysts of *Aiptasia pallida* are composed of monomeric units of a collagenlike protein which are linked by disulfide bonds (Blanquet and Lenhoff 1966). Phelan and Blanquet (1985) showed that the capsule and tubule proteins of *Aiptasia* nematocysts contain about 20% glycine and 28–33% proline-hydroxyproline. They supported the inclusion of the proteins of the capsule and tubule of nematocysts in the collagen class. The capsule wall of most cnidae consists of two layers, and these layers are fibrous and in crossed helicoidal arrangement (Skaer and Picken 1965; Robson 1973).

Blanquet (1968) analyzed the capsule contents of microbasic mastigophore nematocysts isolated from acontial filaments of *Aiptasia pallida*. The nematocyst contents are composed of two proteins, whose molecular weights are 31 and 132 kDa. Only the protein of higher molecular weight has toxic activity. He found that 76.3% of the nematocyst contents are glutamic acids and suggested that either both protein fractions have large segments composed of glutamic acid or one may be almost exclusively polyglutamic acid. He also observed that nematocyst contents discharged into distilled water precipitate between pH 2.6 and 3.8 and suggested that an isoelectric point of the nematocyst proteins was in this range. Blanquet (1970) suggested that divalent cations stabilize the acidic proteins by forming calcium or magnesium bridges between neighboring anions. Phelan and Blanquet (1985) confirmed the high percentage of glutamic acid and/or glutamine in whole *Aiptasia* venom. The 31.8 kDa venom protein, however, contains glycine and glutamic acid, which accounts for 80 and 20% of the total, respectively.

More than 80% of the intracapsular fluid substances of *Hydra* nematocysts are acidic proteins of small molecular weight (Weber et al. 1987a). These proteins or peptides are rich in glutamic acid, having an isoelectric point below 3. It has been suggested that the highly anionic proteins may be involved in binding divalent cations such as Ca^{2+} and Mg^{2+} and, therefore, in the discharge mechanism (Weber et al. 1987a). Recently, Weber (1990) demonstrated that poly(γ-glutamic acid)s, which have previously been known to occur only in some bacteria, are the major constituents of *Hydra* nematocysts. The acidic proteins in *Hydra* nematocysts are polymers of different chain lengths, built up of L-glutamic acid monomers that are linked via amide bonds between the γ-carboxyl and α-amino groups of adjacent monomers. The potentiometric titration of crude nematocyst extract revealed 2.80 μmol of titrable groups/mg dry nematocyst. Weber (1989) substituted cations of isolated nematocysts with ammonium and then assayed the incorporated NH_4^+. He found that 100% NH_4^+ uptake corresponds to 2.77 μmol/mg dry nematocyst. Thus, the number of titrable groups on the polyanions closely matches the concentrations of cations in the nematocyst. The in situ concentration of the intracapsular glutamic acid monomer is estimated to be as high as 2 *M* (Weber 1990). Weber (1991) showed that nematocysts of other cnidarians also consist of poly(γ-glutamic acid)s with various degrees of polymerization. These polyanions generate a highly asymmetric distribution of ions across the capsule wall in accordance with the Donnan equilibrium and are, in cooperation with corresponding cations, responsible for the generation and regulation of an internal osmotic pressure (Weber 1989).

6 The Donnan Equilibrium and Ionic Mechanism of Nematocyst Discharge

6.1 Nematocysts as Donnan Equilibrium Systems

Nematocysts contain polyanions and associated cations. The capsule wall is permeable to water and small ions but impermeable to the polyanions. Hence, nematocysts can be considered as Donnan equilibrium systems (Weber 1989). Weber (1989) compared the affinity of various cations for the capsular content by examining salt concentrations which inhibit a massive cooperative binding of the metachromatic dye acridine orange to the contents of isolated nematocysts. He found that divalent cations have a 15–50 times higher apparent affinity than monovalent cations for the polyanionic capsular contents. However, no difference was noted in the affinity for the capsular contents among divalent cations or among monovalent cations. Weber (1989) suggested that most of the cations in nematocysts are not selectively bound to the polyanionic capsular contents but that they occur as fully hydrated ions in the capsule interior, thus generating a colloid osmotic pressure.

The concentration of glutamic acid monomers within hydra nematocysts is estimated to be about $2M$ (Weber 1990a), and almost equivalent concentrations of cations are found in isolated (Weber et al. 1987b; Weber 1989) and in situ nematocysts (Tardent et al. 1990) of *Hydra*. The average degree of polymerization of poly(γ-glutamic acid)s is estimated to be 10–50, suggesting that the polymer concentration is relatively low, ranging from 0.04 to $0.2M$. Thus, cations associated with the polyanions rather than the polyanions themselves contribute much to the high osmotic pressure of the intracapsular fluid.

Weber (1989) examined the volumetric behavior of isolated stenoteles in different salt solutions. For the monovalent cations Na^+ and K^+, the capsule swells from 100 to 115% of its original volume when the external salt concentration is lowered from $3M$ to $1 mM$ at pH 9 (Fig. 6). When the cations are divalent, the capsule swells similarly but at lower absolute values. Weber (1989) performed simulation studies on the behavior of the polyanions and associated cations of nematocysts immersed in salt solutions of various concentrations. He developed an equation for calculating the concentration of free cations in the nematocyst following the methods used for calculating the osmotic properties of ionic gels (Ohmine and Tanaka 1982). The model fits the observed osmolarity differences for stenoteles provided that the affinity of the polyanions for the cations is very low and that the external pH is 9.

When nematocysts stained in 0.1% toluidine blue were induced to discharge in $5 mM$ EGTA, clear water flowed into the capsule. The stained intracapsular fluid was not diluted with the water that came from outside but was extruded through the tip of the everted tubule (Hidaka 1992). The capsule of discharged nematocysts became completely clear or a few drops of stained fluid remained (Fig. 7). The polyanionic capsular contents of nematocysts bind metachromatic dyes such as toluidine blue (Weber 1990). Since the stained intracapsular contents did not mix with water, they cannot be osmotically active. This supports the idea that cations associated with the polyanions contribute to the high internal osmotic pressure rather than the polyanionic capsular contents themselves. If this is the case, the venom can be injected to the target tissue without much dilution with water coming from outside during the discharge.

6.2 Ionic Mechanism of Nematocyst Discharge

It is unlikely that depolymerization of poly(γ-glutamic acid)s occurs to increase the osmolarity of the intracapsular fluid at the time of discharge, since the activity of polyglutamyl synthetase which polymerizes poly(γ-glutamic acid)s in bacteria has not been found in hydra in preliminary

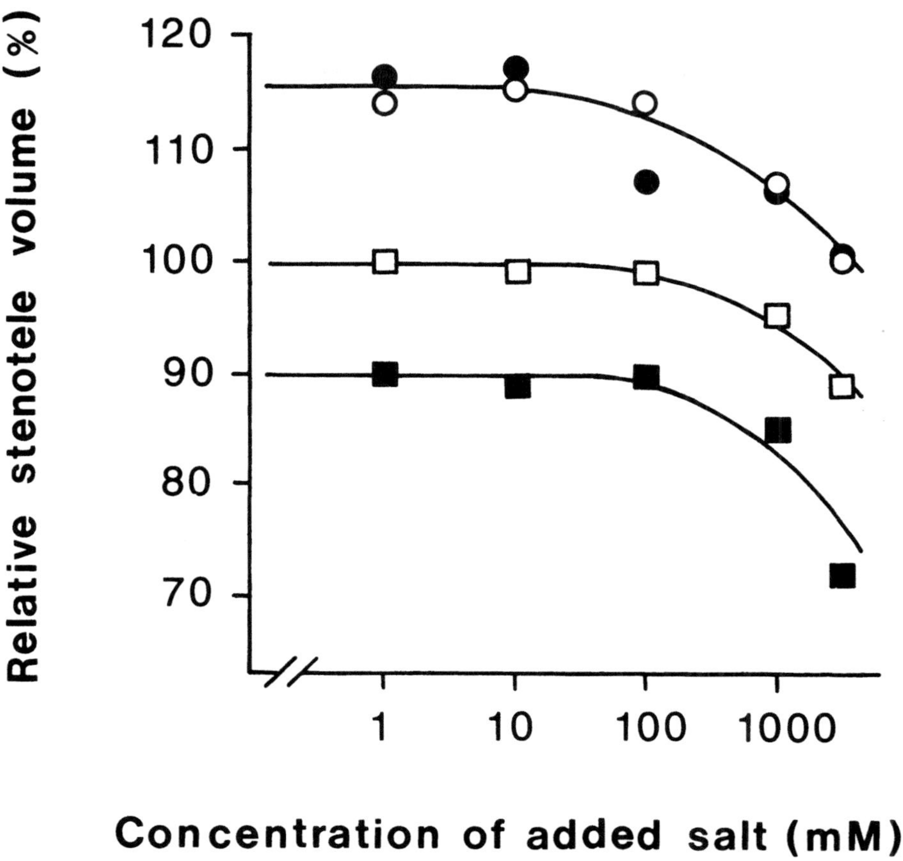

Fig. 6. Volumetric behavior of isolated hydra stenoteles in different salt solutions. The volume of stenoteles loaded with Na^+ (○), K^+ (●), Mg^{2+} (□), and Ca^{2+} (■) were measured in 0.001–3 *M* NaCl, KCl, $MgCl_2$, and $CaCl_2$, respectively. (Weber 1989)

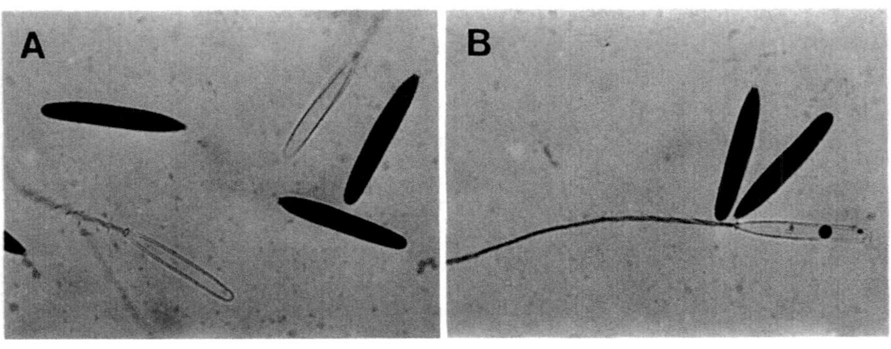

Fig. 7A,B. Photomicrographs of nematocysts which were stained in 0.1% toluidine blue and then immersed in 5 m*M* EGTA. Discharged nematocysts became completely clear (**A**) or a few drops of the stained intracapsular fluid remained in the capsule (**B**), while undischarged nematocysts remained darkly stained

experiments (Weber 1990). It follows that if the osmotic pressure of the intracapsular fluid increases at the time of discharge, it might be due to changes in the ionic composition of the intracapsular fluid or of the medium surrounding the nematocyst.

In situ nematocysts are surrounded by the nematocyst membrane, which may act as a permeability barrier between the cytoplasm and the nematocyst (Lubbock et al. 1981). This means that in situ nematocysts are surrounded by a solution which is in equilibrium with the intracapsular fluid and, therefore, contains high concentrations of ions. Stenoteles loaded with Na^+ or K^+ swell up to 115% and tend to discharge when the external salt concentration is lowered (Weber 1989). In situ stenoteles might contain high concentrations of K (Zierold et al. 1989; Tardent et al. 1990a; Gerke et al. 1991) and may behave as isolated stenoteles loaded with K^+. If this is the case, the osmotic pressure difference across the capsule wall increases when nematocysts are exposed to a dilute salt solution. This occurs when the nematocysts of hydra come into contact with fresh water as a result of exocytosis and may lead to their discharge (Fig. 8A).

The discharge of nematocysts in marine cnidarians cannot be accounted for by the above mechanism, since the external salt concentration is high. Lubbock et al. (1981) found high concentrations of calcium in undischarged nematocysts of sea anemones using the same technique as Zierold et al. (1989) and Tardent et al. (1990a). If the intracapsular Ca^{2+} is replaced by the major cation of sea water, Na^+, when nematocysts are exposed to sea water, the internal osmotic pressure of the nematocysts would increase (Fig. 7B). It is a tempting idea that the Ca^{2+}-Na^+ exchange increases the internal osmotic pressure, leading to discharge. This hypothesis is consistent with the observation made by Lubbock et al. (1981) that an influx of sodium and chloride accompanies the efflux of calcium at the time of discharge. To test this hypothesis I examined the volumetric behavior of isolated nematocysts in various concentrations of salt solutions. When nematocysts isolated from the sea anemone *Calliactis polypus* in artificial sea water (ASW) were immersed in successively lower concentrations of salts, the volume of the nematocysts increased if the cations were divalent but decreased if the cations were monovalent. On lowering the concentration of cations from 1 *M* to 1 m*M*, the volume of the capsule changed as follows: Ca^{2+} from 86 to 95% of the original volume in ASW, Mg^{2+} 96 to 99%, Na^+ 109 to 97%, and K^+ 106 to 98% (Hidaka unpubl. observ.). The observation that isolated nematocysts shrank in concentrated solutions of divalent cations but swelled in concentrated solutions of monovalent cations supports the idea that an exchange of intracapsular Ca^{2+} with external Na^+ increases the internal osmotic pressure.

The responses of isolated nematocysts to dilute or concentrated salt solutions are markedly different depending on the media in which the nematocysts are isolated (Hidaka and Mariscal 1988; McKay and Anderson 1988a; Greenwood and Garrity, 1990). For instance, nematocysts isolated

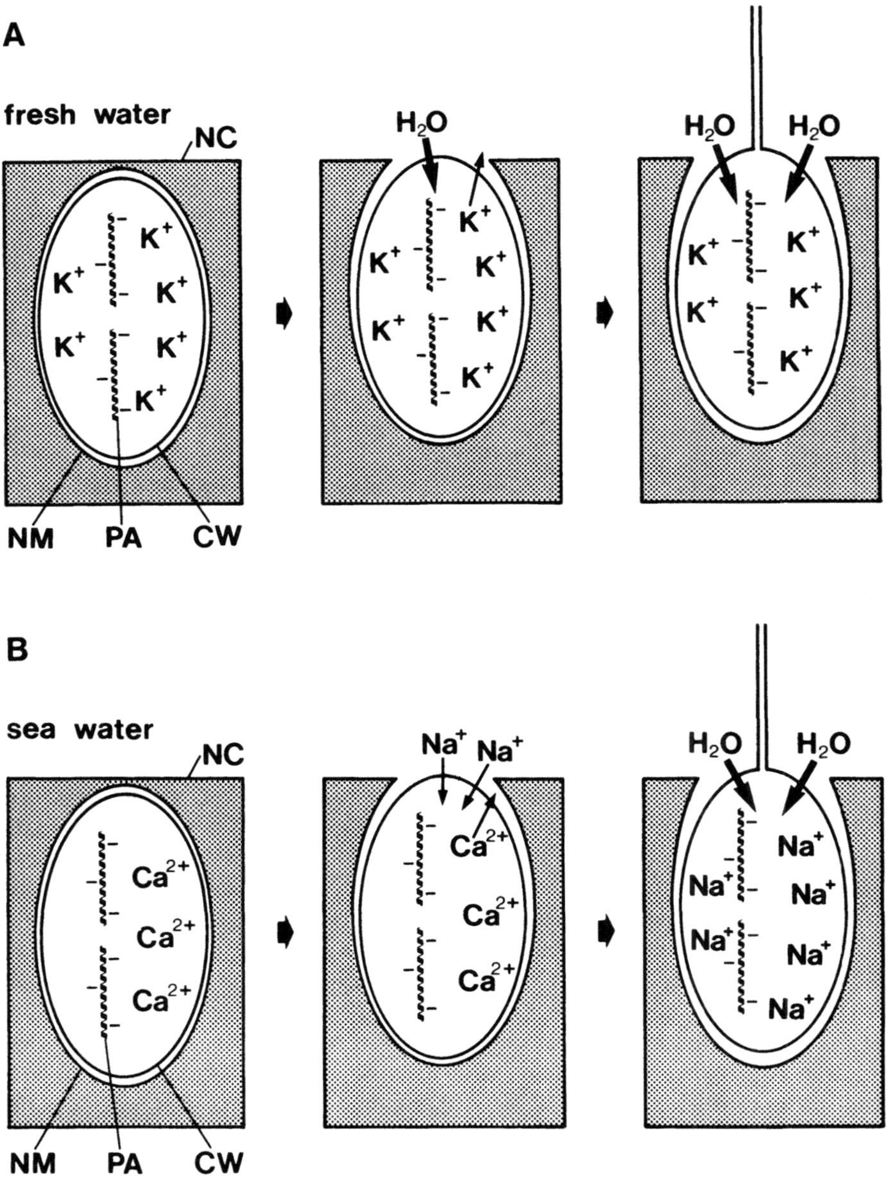

Fig. 8.A,B. Schematic diagram of the process of nematocyst discharge based on the assumption that nematocysts behave as Donnan-potential-dominated systems as suggested by Weber (1989). **A** When nematocysts of hydra are exposed to fresh water, the osmotic pressure difference across the capsule wall increases, resulting in discharge. **B** If Ca^{2+} within the capsule is replaced by Na^{+} when nematocysts of marine cnidarians are exposed to sea water, the internal osmotic pressure would increase and may result in discharge. *CW* Capsule wall; *NC* nematocyte; *NM* nematocyst membrane; *PA* polyanion

in distilled water or 1 *M* glycerol discharge when immersed in sea water or isotonic salt solutions (Yanagita 1959b; Salleo et al. 1984b). In contrast, nematocysts isolated in ASW or 1 *M* sodium citrate discharge when immersed in distilled water or dilute salt solutions (Blanquet 1970; Hidaka and Mariscal 1988). If nematocysts isolated in distilled water or 1 *M* glycerol retain the naturally occurring cation Ca^{2+}, the Ca^{2+}-Na^{+} exchange hypothesis predicts that the nematocysts tend to discharge in solutions containing high concentrations of monovalent cations. On the other hand, if the naturally occurring cation is replaced by Na^{+} in nematocysts isolated in ASW or 1 *M* sodium citrate, the nematocysts tend to discharge in fresh water and dilute salt solutions like hydra nematocysts loaded with Na^{+} or K^{+}. Thus, the varying responsiveness of nematocysts isolated in different media can be explained if the nematocyst cations are replaced by externally applied ones and if the Ca^{2+}-Na^{+} exchange generates a high osmotic pressure sufficient to elicit nematocyst discharge.

A preliminary, semiquantitative, X-ray microanalysis showed that major elements of *Calliactis* nematocysts incubated in ASW were, besides sulfur, calcium and magnesium (Hidaka unpubl. observ.). Only a fraction of calcium was replaced by Na^{+} when nematocysts were incubated in 1 *M* NaCl, contrary to the observation with hydra stenoteles (Weber 1989). Even this partial exchange would increase the osmolarity of the intracapsular fluid by 16% if in situ nematocysts contain exclusively Ca^{2+}. This value could be underestimated since cations with a low affinity for the polyanions might be lost during washing in distilled water and make the apparent percentage of Na^{+} in the nematocysts low.

The volume of nematocysts immersed in 1 *M* $CaCl_2$ was smaller than the volume of those immersed in 1 *M* $MgCl_2$ (Hidaka unpubl. observ.). This suggests that Ca^{2+}-Mg^{2+} exchange would increase the internal osmotic pressure and that Ca^{2+} selectively binds to the polyanions to reduce the osmolarity of the intracapsular fluid. Ca^{2+} may reduce the internal osmotic pressure by cross-linking the polyanions as suggested by Lubbock et al. (1981) or simply by binding to negatively charged groups on the polyanions. If the number of negative charges on the polyanions decreases, the number of free cations associated with the polyanions would decrease, reducing the internal osmotic pressure. Thus, the osmotic pressure of the intracapsular fluid is determined not only by a Donnan equilibrium but also by the selective binding of Ca^{2+} to the polyanions. To understand which component plays the greater role, it is necessary to know the osmotic pressure and the ionic composition of the nematocysts immersed in various salt solutions.

The problem of the Ca^{2+}-Na^{+} exchange hypothesis is that the exposure of nematocysts to sea water alone does not elicit a discharge of the nematocysts, as shown by the fact that undischarged nematocysts can be isolated in sea water (Glaser and Sparrow 1909; Hidaka and Mariscal 1988). This suggests that Ca^{2+}-Na^{+} exchange, if it occurs at all, does not increase the internal osmotic pressure to a extent sufficient to trigger discharge. Furthermore, isolated nematocysts can be induced to discharge in solutions which contain only a small amount of monovalent cations by reagents such as disulfide reducing agents and calcium chelators.

Tardent et al. (1990b) examined the concentration of metallic cations in nematocysts of various marine cnidarians by means of X-ray microanalysis of rapidly frozen cryosections. The predominant cation of in situ nematocysts is different among species and among nematocyst types and is either Ca^{2+}, Mg^{2+} or K^+. The dominant cation of the tentacular and acontial nematocysts of *Calliactis parasitica* is unexpectedly K^+. An increase in the internal osmotic pressure via cation substitution cannot account for the discharge of the nematocysts which contain predominantly K^+. In these nematocysts, the osmotic pressure difference across the capsule wall would decrease via cation substitution when the nematocysts are exposed to sea water. These nematocysts might discharge when an instantaneous increase in the osmotic pressure difference exceeds a certain limit when the nematocysts come into contact with sea water. Another possibility is that the "stopper" is weakened at the time of discharge by some unknown mechanism. It is still unknown whether discharge of nematocysts is a physicochemical response to changes in the ionic composition of the surrounding medium or involves biochemical modification of critical structural components of the nematocysts.

7 Cellular Control of Nematocyst Discharge

The nematocysts in situ within the body surfaces respond to mild stimuli such as a slight mechanical contact in the presence of appropriate chemosensitizer(s), while rather violent treatment involving, for example, extreme pH values, proteolytic enzymes, and disulfide reducing agents is necessary to elicit the discharge of isolated nematocysts (e.g. Yanagita 1960a; Thorington and Hessinger 1988a). Electric stimuli induce the discharge of in situ nematocysts but are not effective for eliciting the discharge of isolated nematocysts (Glaser and Sparrow 1909; Yanagita 1960a). These observations suggest that cnidocytes and accessory cells may play an important role in transducing the external stimuli to the nematocyst. Agents that induce discharge of isolated nematocysts are not necessarily involved in the in situ mechanism of discharge. To understand the latter, it is necessary to know the role of the cnidocytes and accessory cells in nematocyst discharge.

7.1 Receptors of Nematocytes

The ciliary apparatus on the apical surface of nematocytes is considered as the receptor responsible for nematocyst discharge (Mariscal 1974; Mariscal et al. 1978). The ciliary apparatus consists of a single cilium surrounded by microvilli and stereocilia. The cilium exhibits a 9 + 2 microtubule arrangement in nematocytes from anthozoans (Westfall 1965; Watson and Mariscal 1985). Nematocytes from hydrozoans and scyphozoans have a modified cilium called a cnidocil, which has a 9 + 4 microtubule arrangement in *Craspedacusta* (Hausmann and Holstein 1985), nine doublets and a central, dense core of microtubules in *Gonionemus* (Westfall 1970a) and *Hydra* (Slautterback 1967), and numerous, longitudinally arranged, doublet and singlet microtubules in *Physalia* (Cormier and Hessinger 1980a). The entire ciliary apparatus is called the cnidocil apparatus in hydrozoan and scyphozoan nematocytes and the ciliary cone in anthozoan nematocytes. While the cnidocil apparatus originates solely from cnidocytes (Cormier and Hessinger 1980a; Hausmann and Holstein 1985; Rifkin and Endean 1988), the ciliary cone is a multicellular structure; a cilium and microvilli originate from the nematocyte, but the stereocilia originate from two to five supporting cells surrounding the nematocyte (Bigger 1982; Watson and Hessinger 1988). Spirocytes do not have the ciliary apparatus but are associated with two kinds of microvilli. Small microvilli of the spirocyte encircle the spirocyst tip, and larger microvilli from adjacent cells form an outer ring (Mariscal et al. 1976; Bigger 1982).

The in situ discharge of cnidae requires appropriate mechanical and chemical stimuli (Parker and van Alstyne 1932; Pantin 1942). This suggests that chemical stimuli make the cnidocytes more sensitive to tactile stimuli. The responses of tentacles and acontia of sea anemones to various chemicals have been investigated in an attempt to find agents that lower the threshold of the cnidocyte to mechanical contact (Pantin 1942; Yanagita 1960a; Lubbock 1979). At least two groups of substances have been identified which chemosensitize in situ cnidocytes in the sea anemone *Aiptasia pallida* (Giebel et al. 1988; Thorington and Hessinger 1988a,b, 1990). The chemosensitizing substances are *N*-acetylated sugars and a wide variety of amino compounds. These chemicals act via at least two classes of chemoreceptors to increase the sensitivity of cnidocytes to mechanical stimuli. Watson and Hessinger (1988, 1989a) studied the sugar receptor on feeding tentacles of *Haliplanella luciae* using colloidal gold coated with the glycoprotein, bovine submaxillary mucin. They found that over 99% of the mucin-gold particles appeared on or within the supporting cells. Mucin-gold sensitizes cnidocytes to discharge nematocysts in a dose-response manner. The number of mucin-gold particles appearing on the surface of supporting cells changes over time, and such changes correlate with the time course of cnidocyte responsiveness (Watson and Hessinger 1989a). They suggested that the discharge of nematocysts by cnidocytes may be regulated by the number of receptor-ligand complexes at the surface of the adjacent supporting cells and that cnidocytes and adjacent supporting cells form receptor-effector complexes.

Cnidocyte-supporting cell complexes of the tentacle of the sea anemone *Haliplanella luciae* discharged nematocysts preferentially into test-probes vibrating at 30, 50–55, and 75 Hz. However, in sea water containing mucin, the cnidocyte-supporting cell complexes discharged nematocysts maximally into probes vibrating at 5, 15, 30, and 40 Hz, frequencies that match those produced by swimming prey (Watson and Hessinger 1989b). The stereocilia in cnidocyte mechanoreceptors elongate immediately upon exposure to *N*-acetylneuraminic acid (Watson and Hessinger 1990). Watson and Hessinger (1989b, 1990) suggested that chemoreceptors for *N*-acetylated sugars and mucin tune the cnidocyte mechanoreceptors to frequencies that match the movement of prey by altering the length of the sterocilia.

7.2 Fibrillar Systems Surrounding Nematocysts

The nematocyst capsule in tentacles of *Physalia*, *Chironex*, and *Hydra* is surrounded by a fibrillar system composed of a network of microtubules and microfilaments (Cormier and Hessinger 1980a; Rifkin and Endean 1988; Stidwill and Honegger 1989). The microtubules are arranged in parallel arrays, forming a basketlike structure around the mature nematocyst. The fibrillar system may serve to support the nematocyst in a proper orientation in the nematocyte and to retain the discharged nematocyst within the epithelium. Watson and Mariscal (1985) also found a microtubule array surrounding the capsule tip of nematocysts in the catch tentacles of the sea anemone *Haliplanella luciae*. They suggested that the microtubule array and intermembrane particles between the plasma membrane and the membrane that covers the nematocyst capsule help to hold the nematocyst capsule at the cell surface until discharge. Contrary to the observation by Lubbock et al. (1981), mature nematocysts of acrorhagi and catch tentacles of sea anemones are already located at the plasma membrane (Bigger 1982; Watson and Mariscal 1985). The fibrillar system may also serve as a regulatory element of discharge or a mediator of a stimulus (Cormier and Hessinger 1980a; Stidwill and Honegger 1989).

Cormier and Hessinger (1980b) found that the stereocilia which surround the cnidocil of *Physalia* nematocytes contain closely packed, longitudinally arranged microfilaments. Wood and Novak (1982) observed that the basal portion of the cnidocil supportive apparatus of *Hydra* binds N-[7-nitrobenz-2-oxa-1,3-diazol-4-yl]phallacidin (NBD-phallacidin), indicating the presence of actin. Recently, Stidwill et al. (1988) observed that bundles of actin filaments of uniform polarity surround the cnidocil in the apical part of hydra nematocytes. The actin filament bundles make contact with other cytoskeletal elements and may serve as a transducer of the mechanical stimuli from the cnidocil to the cyst (Stidwill et al. 1988; Hausmann and Holstein 1985).

In the cubozoan *Chironex fleckeri*, the fibrillar basket and fibrous bundles which surround a nematocyst are connected to the mesoglea by radially oriented fibers (Rifkin and Endean 1988). Radial fibers are also present in the accessory cells which surround the nematocyte. They suggested that the radial fibers are contractile. When the radial fibers in the nematocyte contract, the fibrillar system surrounding the nematocyst is retracted basally within the cell. The microvilli and the cnidocil, which are attached to the fibrous basket, are also pulled basally into an invagination at the apical end of the nematocyte. Retraction of the accessory cells surrounding the nematocyte in turn brings the nematocyte in a fire-ready position, allowing the triggering apparatus to protrude from the cell. Similar changes in the shape of nematocytes occur during the expansion and contraction of acrorhagi (Bigger 1982). While the apical surface of a nematocyte and microvilli are folded into an invagination of the nematocyte in an unexpanded acrorhagus, they are exposed to the environment in an expanded acrorhagus. Since acrorhagial nematocyst discharge requires prior acrorhagial expansion, Bigger (1982) suggested that such a change in shape could be responsible for the threshold change of nematocyst discharge.

7.3 Responses of Isolated Cnidocytes and In Situ Nematocysts

McKay and Anderson (1988a,b) and Anderson and McKay (1987) isolated cnidocytes from various cnidarians using papain digestion followed by density centrifugation in a Percoll-containing medium. The isolated cnidocytes appear to have an intact plasma membrane and support transmembrane ionic currents and negative resting potentials. They examined whether excitation of the cnidocyte led to discharge of the cnida. Cnida discharge is not triggered by imposed changes in membrane potential nor by selective blockade of the ionic current present in the cell (Anderson and McKay 1987; McKay and Anderson 1988a). Isolated cnidocytes from *Cladonema* tentacles produce Na^+-driven action potentials, while cnidocytes isolated from *Chrysaora* and in situ cnidocytes of *Physalia* do not produce action potentials (Anderson and McKay 1987). Stenoteles contained in nematocytes isolated from *Cladonema* do not discharge when the action potentials occur in the nematocytes. Nematocytes isolated from *Cladonema* and *Chrysaora* sometimes discharge, but their discharge is not correlated with any obvious electrophysiological event. Cnidocytes in situ in tentacles of *Physalia* do not discharge in response to changes in membrane potential (Anderson and McKay 1987). They are, however, different from other cnidocytes in that they do not display delayed rectification (McKay and Anderson 1988a).

The above observations appear to contradict the fact that cnidae in situ can be induced to discharge by electric stimuli or high potassium concentrations (e.g., Yanagita 1960a; Rifkin 1982). Anderson and McKay (1987) suggested that the magnitudes of the electrical shocks used may be nonphysiological and that the discharge may merely represent breakdown of the cell or cnida membrane. However, it is also possible that excitation of the isolated cnidocyte did not lead to discharge of the cnida because its transduction system was impaired during the isolation procedure. While 0.5 *M* KCl invariably elicits discharge of in situ cnidocytes, it elicits discharge of isolated spirocytes only occasionally (McKay and Anderson 1988b). The cnidocil apparatus was observed only sometimes or on a few isolated cnidocytes (Anderson and McKay 1987; McKay and Anderson 1988b). The absence of supporting cells may also affect the transduction system (Watson and Hessinger 1989a). Thus, it seems too early to conclude that cnida discharge is not coupled with cnidocyte excitation.

Santoro and Salleo (1991) reported that in the presence of 10 m*M* Ca^{2+}, 0.553 *M* KCl produces a partial discharge of in situ nematocysts and extrusion of undischarged nematocysts in acontia of *Aiptasia mutabilis*. Cnidocytes in tentacles of *Anthopleura elegantissima* discharge in response to 0.5 *M* KCl when the tentacles have been immersed in ASW, Na-free ASW, or Mg-free ASW, but not when they have been immersed in Ca-free ASW (McKay and Anderson 1989b). Thus, the discharge of in situ nematocysts induced by K^+ requires Ca^{2+}. Santoro and Salleo (1991) suggested that a high K^+ concentration causes depolarization in nematocytes or accessory cells, inducing an increase in the Ca^{2+} conductance, which, in turn, leads to discharge. The discharge responses of cnidae in *Anemonia* tentacles and in *Haliplanella* acontia to high K^+ concentrations and electric stimuli were abolished by magnesium sea water (Pantin 1942; Yanagita 1960a).

7.4 Nematocyte Innervation

The behavior of some cnidarians suggests that cnida discharge is controlled by the animal (Mariscal 1974; Thorington and Hessinger 1988a). The tentacles of the swimming sea anemone *Stomphia coccinea* do not adhere to food material when the pedal disc is detached from the substratum, that is, during swimming (Ross and Satton 1964). When the symbiotic sea anemones *Calliactis parasitica* are already settled on shells, fewer tentacles adhere to a second shell than when they are attached to other substrata (Davenport et al. 1961). However, once tentacles of the anemones *Calliactis tricolor* and *C. parasitica* adhere to a presented shell, the number of cnidae discharged to the shell is independent of the nature of the substratum to which the anemones are attached (Mariscal 1972; McFarlane and Shelton 1975). Conklin and Mariscal (1976) showed that tactile stimulation of the body column of the anemone *Anemonia sargassensis* increases the discharge of tentacle cnidae to coverslips coated with food extract. During locomotion, a hydra uses its tentacles to cling to the substratum. If unattached tentacles of a locomoting hydra are stroked with a glass rod, a significantly smaller number of atrichous isorhizas are discharged to the glass rod than when tentacles of a sedentary hydra are stroked (Kass-Simon 1988). These observations suggest that cnidarians can inhibit or activate the discharge of cnidae presumably via the nervous system, though cnidarians may also control cnida discharge indirectly by altering their behavior.

Cnida discharge is also influenced by the physiological state of the animal. An inhibition of nematocyst discharge following feeding has been reported in hydra (e.g., Smith et al. 1974), the sea anemone *Calliactis tricolor* (Sandberg et al. 1971; Mariscal 1973), and the colonial hydroid *Halocordyle disticha* (Clark and Cook 1986). The inhibition of nematocyst discharge may be due to metabolites from the prey (Smith et al. 1974; Grosvenor and Kass-Simon 1987) or to soluble factors released from a feeding animal, probably from discharged nematocysts themselves (Ruch and Cook 1984; Clark and Cook 1986). If an individual tentacle is isolated and *Artemia* homogenate pipetted onto it, the segment of the tentacle which was exposed to the homogenate no longer captures or kills *Artemia* (Kass-Simon 1988). Grosvenor and Kass-Simon (1987) and Kass-Simon (1988) suggested that the receptors for these inhibitory substances are found on the surface of the hydra tentacle and may be associated with the nematocyte directly or with some other component of the battery cell complex. Thus, the inhibition of nematocyst discharge by food molecules in hydra might be local, though this does not exclude the possibility that the inhibition of nematocyst discharge by satiated hydra is controlled also by the nervous system.

Ultrastructural studies show synaptical contact on nematocytes, suggesting a possible role of the nervous system in the control of nematocyst discharge in hydra (Westfall et al. 1971; Westfall 1973; Westfall and Kinnamon 1978, 1984; Hufnagel and Kass-Simon 1988) and in *Gonionemus* (Westfall 1970b). The ectoderm of the distal tentacle of hydra is composed exclusively of battery cell complexes, each consisting of a battery cell enveloping separately nested nematocytes, a ganglion cell, and, though not always, a sensory cell (Hufnagel et al. 1985). Ganglion cells synapse onto nematocytes as well as onto other cells in a battery cell complex (Hufnagel and Kass-Simon 1988). Neurites of unknown origin also synapse onto nematocytes as well as onto other cells and neurites in a battery cell complex. Neurites can extend from a battery cell complex to a neighboring battery cell complex. Yu et al. (1985) also showed that an individual neuron can contact up to five different batteries of nematocysts and that different neurons can innervate the same battery.

As compared with morphological studies, there are few pharmacological or electrophysiological studies that support the nervous control of nematocyst discharge. Lentz and Barrnett (1962) placed hydra in a culture medium containing neurotransmitters, stroked the tentacles, and then counted the number of discharged nematocysts with a light microscope. Acetylcholine produces a considerable discharge, especially of stenoteles, and this is augmented by eserine. The effect of acetylcholine is blocked by hexamethonium and D-tubocurarine, blockers of nicotinic receptors. However, atropine, a blocker of muscarinic receptors, does not block the effect of acetylcholine but causes a discharge. Lentz and Barrnett (1961) have identified cholinesterase activity in the grapelike clusters of small round bodies which appeared to be attached to the surface of the nematocytes of hydra. Monoamines such as epinephrine, norepinephrine, 5-hydroxytryptamine, and histamine augment discharge, especially of isorhizas. Lentz and Barrnett (1962) suggested that the nervous system of hydra plays a significant role in nematocyst discharge and that neural control of the stenoteles might be different from that of the isorhizas.

When an acrorhagial tip of the sea anemone *Anthopleura elegantissima* is touched with an allogeneic tentacle, a massive simultaneous discharge of nematocysts, including those remote from the point of contact, occurs (Lubbock and Shelton 1981). The discharge is accompanied by a triphasic spike up to 50 μV recorded extracellularly. A variable number of much smaller impulses up to 10 μV was also recorded. The amplitude of the impulses decreases rapidly as the distance from the acrorhagial tip increases. The small impulses usually precede the spike and nematocyst discharge but can be evoked without the subsequent spike or nematocyst discharge if the contact with the allogeneic material is brief. Lubbock and Shelton (1981) suggested that specific molecules on the surface of the allogeneic tissue bind to surface receptors on cells at the tip of the acrorhagus, inducing electrical excitation; local electric potentials summate up to a point at which massive nematocyst discharge accompanied by a spike occurs.

Aerne et al. (1991) constructed *Hydra* that lack nerve cells but whose tentacles contain nematocytes, and demonstrated that nematocysts in nerve-free *Hydra* can be induced to discharge by mechanical or electrical stimulation. They suggested that at least in *Hydra* nematocytes are capable of functioning as autonomous receptor-effector units independent of the nervous system. This, however, does not exclude the possibility that the nervous system of *Hydra* modifies the response of nematocytes to external stimuli.

8 Conclusions

Recent studies suggest that nematocysts can be considered as Donnan-potential-dominated osmotic systems. Cations associated with polyanions in the capsule rather than the polyanions themselves contribute to the high intracapsular osmotic pressure. This view contrasts with the previous view of nematocyst discharge: Polyanionic capsular contents are cross-linked by Ca^{2+}, and a loss of Ca^{2+} dissociates macromolecules into osmotically active small molecules. The responses of isolated nematocysts to various agents should be reinterpreted from this new viewpoint.

The capsule wall of nematocysts is expanded due to a high internal osmotic pressure and stores elastic energy. When nematocysts are exposed to the external solution as a result of exocytosis, the osmotic pressure difference across the capsule wall might increase to a point at which the "stopper" ruptures, allowing the tubule to evert. It is still not clear how the internal osmotic pressure increases. The possibility cannot be ruled out that the "stopper" is softened at the time of discharge by some unknown agent. Nematocytes and accessory cells play an important role in transducing external stimuli to nematocysts. The in situ mechanism of nematocyst discharge and how external stimuli are transduced to nematocysts are still incompletely understood.

Acknowledgements

I would like to thank Drs. P.G. Greenwood, T. Motokawa, and M.A. Williams for helpful comments on the manuscript and Prof. Keiichi Takahashi for giving me the opportunity to write this review.

References

Aerne BL, Stidwill RP, Tardent P (1991) Nematocysts discharge in *Hydra* does not require the presence of nerve cells. J Exp Zool 258:137–141

Anderson PAV, McKay MC (1987) The electrophysiology of cnidocytes. J Exp Biol 133:215–230

Bigger CH (1982) The cellular basis of the aggressive acrorhagial response of sea anemones. J Morphol 173:259–278

Blanquet R (1968) Properties and composition of the nematocyst toxin of the sea anemone, *Aiptasia pallida*. Comp Biochem Physiol 25:893–902

Blanquet R (1970) Ionic effects on discharge of the isolated and in situ nematocysts of the sea anemone, *Aiptasia pallida*: a possible role of calcium. Comp Biochem Physiol 35:451–461

Blanquet R, Lenhoff HM (1966) A disulfide-linked collagenous protein of nematocyst capsules. Science 154:152–153

Carré D (1980) Hypothesis on the mechanism of cnidocyst discharge. Eur J Cell Biol 20:265–271

Clark SD, Cook CB (1986) Inhibition of nematocyst discharge during feeding in the colonial hydroid *Halocordyle disticha* (= *Pennaria tiarella*): the role of previous prey-killing. Biol Bull 171:405–416

Conklin EJ, Mariscal RN (1976) Increase in nematocyst and spirocyst discharge in a sea anemone in response to mechanical stimulation. In: Mackie GO (ed) Coelenterate ecology and behaviour. Plenum Press, New York, pp 549–558

Cormier SM, Hessinger DA (1980a) Cellular basis for tentacle adherence in the Portuguese man-of-war (*Physalia physalis*). Tissue Cell 12:713–721

Cormier SM, Hessinger DA (1980b) Cnidocil apparatus: sensory receptor of *Physalia* nematocytes. J Ultrastruct Res 72:13–19

Davenport D, Ross DM, Sutton L (1961) The remote control of nematocyst discharge in the attachment of *Calliactis parasitica* to shells of hermit crabs. Vie Milieu 12:197–209

Gerke I, Zierold K, Weber J, Tardent P (1991) The spatial distribution of cations in nematocytes of *Hydra vulgaris*. In: Williams RB, Cornelius PFS, Hughes RG, Robson EA (eds) Coelenterate biology: recent research on cnidaria and ctenophora. Kluwer, Dordrecht, pp 661–669

Giebel GEM, Thorington GU, Lim RY, Hessinger DA (1988) Control of cnida discharge: II. Microbasic p-mastigophore nematocysts are regulated by two classes of chemoreceptors. Biol Bull 175:132–136
Glaser OC, Sparrow CM (1909) The physiology of nematocysts. J Exp Zool 6:361–382
Godknecht A, Tardent P (1988) Discharge and mode of action of the tentacular nematocysts of *Anemonia sulcata* (Anthozoa: Cnidaria). Mar Biol 100:83–92
Greenwood PG, Garrity LK (1991) Discharge of nematocysts isolated from aeolid nudibranchs. In: Williams RB, Cornelius PFS, Hughes RG, Robson EA (eds) Coelenterate biology: recent research on cnidaria and ctenophora. Kluwer, Dordrecht, pp 671–677
Grosvenor W, Kass-Simon G (1987) Feeding behavior in *Hydra*. I. Effects of *Artemia* homogenate on nematocyst discharge. Biol Bull 173:527–538
Gupta BL, Hall TA (1984) Role of high concentration of Ca, Cu, and Zn in the maturation and discharge in situ of sea anemone nematocysts as shown by X-ray microanalysis of cryosections. In: Bolis L, Zadunaisky J, Gilles R (eds) Toxins, drugs, and pollutants in marine animals. Springer, Berlin Heidelberg New York, pp 77–95
Hausmann K, Holstein T (1985) Sensory receptor with bilateral symmetrical polarity. Naturwissenschaften 72:145–147
Hessinger DA, Lenhoff HM (1973) Assay and properties of the hemolysis activity of pure venom from the nematocysts of the acontia of the sea anemone *Aiptasia pallida*. Arch Biochem Biophys 159:629–638
Hidaka M (1983) Effects of certain physico-chemical agents on the mechanical properties of the catch apparatus of the sea-urchin spine. J Exp Biol 103:15–29
Hidaka M (1990) Effects of calcium on the mechanical properties of the capsule wall of isolated nematocysts. Zool Sci 7:1041
Hidaka M (1992) Effects of Ca^{2+} on the volume of nematocysts isolated from acontia of the sea anemone *Calliactis tricolor*. Comp Biochem Physiol 101A:737–741
Hidaka M, Mariscal RN (1988) Effects of ions on nematocysts isolated from acontia of the sea anemone *Calliactis tricolor* by different methods. J Exp Biol 136:23–34
Holstein T, Tardent P (1984) An ultrahigh-speed analysis of exocytosis: nematocyst discharge. Science 223:830–833
Hufnagel LA, Kass-Simon G (1988) Functional anatomy of nematocyte innervation in battery cell complexes of the *Hydra* tentacle. In: Hessinger DA, Lenhoff HM (eds) The biology of nematocysts. Academic Press, San Diego, pp 519–529
Hufnagel LA, Kass-Simon G, Lyon MK (1985) Functional organization of battery cell complexes in tentacles of *Hydra attenuata*. J Morphol 184:323–341
Kass-Simon G (1988) Towards a neuroethology of nematocyst discharge in the tentacles of hydra. In: Hessinger DA, Lenhoff HM (eds) The biology of nematocysts. Academic Press, San Diego, pp 531–541
Klug M, Weber J, Tardent P (1989) Hemolytic and toxic properties of *Hydra attenuata* nematocysts. Toxicon 27:325–339
Lentz TL, Barrnett RJ (1961) Enzyme histochemistry of hydra. J Exp Zool 147:125–149
Lentz TL, Barrnett RJ (1962) The effect of enzyme substrates and pharmacological agents on nematocyst discharge. J Exp Zool 149:33–38
Lubbock R (1979) Chemical recognition and nematocyte excitation in a sea anemone. J Exp Biol 83:283–292
Lubbock R, Amos WB (1981) Removal of bound calcium from nematocyst contents causes discharge. Nature 290:500–501
Lubbock R, Shelton GAB (1981) Electrical activity following cellular recognition of self and non-self in a sea anemone. Nature 289:59–60
Lubbock R, Gupta BL, Hall TA (1981) Novel role of calcium in exocytosis: mechanism of nematocyst discharge as shown by X-ray microanalysis. Proc Natl Acad Sci USA 78:3624–3628
Mariscal RN (1972) The nature of the adhesion to shells of the symbiotic sea anemone *Calliactis tricolor* (Leseur). J Exp Mar Biol Ecol 8:217–224
Mariscal RN (1973) The control of nematocyst discharge during feeding by sea anemones. Publ Seto Mar Biol Lab 20:695–702

Mariscal RN (1974) Nematocysts. In: Muscatine L, Lenhoff HM (eds) Coelenterate biology: reviews and new perspectives. Academic Press, London, pp 129–178
Mariscal RN (1980) The elemental composition of nematocysts as determined by X-ray microanalysis. In: Tardent P, Tardent R (eds) Developmental and cellular biology of coelenterates. Elsevier/North-Holland Biomedical Press, Amsterdam, pp 337–342
Mariscal RN (1984) Cnidaria: Cnidae. In: Bereiter-Hahn J, Matoltsy AG, Richards KS (eds) Biology of the integument, vol 1. Invertebrates. Springer, Berlin Heidelberg New York, pp 57–68
Mariscal RN (1988) X-ray microanalysis and perspectives on the role of calcium and other elements in cnidae. In: Hessinger DA, Lenhoff HM (eds) The biology of nematocysts. Academic Press, San Diego, pp 95–113
Mariscal RN, Bigger CH, McLean RB (1976) The form and function of cnidarian spirocysts. 1. Ultrastructure of the capsule exterior and relationship to the tentacle sensory surface. Cell Tissue Res 168:465–474
Mariscal RN, Conklin EJ, Bigger CH (1978) The putative sensory receptors associated with the cnidae of cnidarians. Scanning Electron Microsc 11:959–966
McFarlane ID, Shelton GAB (1975) The nature of the adhesion of tentacles to shells during shell-climbing in the sea anemone *Calliactis parasitica* (Couch). J Exp Mar Biol Ecol 19:177–186
McKay MC, Anderson PAV (1988a) On the preparation and properties of isolated cnidocytes and cnidae. In: Hessinger DA, Lenhoff HM (eds) The biology of nematocysts. Academic Press, San Diego, pp 273–293
McKay MC, Anderson PAV (1988b) Preparation and properties of cnidocytes from the sea anemone *Anthopleura elegantissima*. Biol Bull 174:47–53
Motokawa T (1984) Connective tissue catch in echinoderms. Biol Rev 59:255–270
Ohmine I, Tanaka T (1982) Salt effects on the phase transition of ionic gels. J Chem Phys 77:5725–5729
Pantin CFA (1942) The excitation of nematocysts. J Exp Biol 19:294–310
Parker GH, Van Alstyne MA (1932) The control and discharge of nematocysts, especially *Metridium* and *Physalia*. J Exp Zool 63:329–344
Phelan MA, Blanquet RS (1985) Characterization of nematocyst proteins from the sea anemones *Aiptasia pallida* and *Pachycerianthus torreyi* (Cnidaria: Anthozoa). Comp Biochem Physiol 81B:661–666
Picken LER, Skaer RJ (1966) A review of researches on nematocysts. Symp Zool Soc Lond 16:19–50
Rifkin JF (1982) Use of electrical stimulation for discharging cnidom components of a species of *Cerianthus* (Anthozoa: Cnidaria). Mar Biol 69:31–36
Rifkin JF, Endean R (1988) Arrangement of accessory cells and nematocytes bearing mastigophores in the tentacles of the cubozoan *Chironex fleckeri*. J Morphol 195: 103–115
Rigby BJ, Hirai N, Spikes JD, Eyring H (1959) The mechanical properties of rat tail tendon. J Gen Physiol 43:265–283
Robson EA (1973) The discharge of nematocysts in relation to properties of the capsule. Publ Seto Mar Biol Lab 20:653–673
Ross DM, Satton L (1964) Inhibition of the swimming response by food and of nematocyst discharge during swimming in the sea anemone *Stomphia coccinea*. J Exp Biol 41:741–757
Ruch RJ, Cook CB (1984) Nematocyst inactivation during feeding in *Hydra littoralis*. J Exp Biol 111:31–42
Salleo A (1984) Discharge mechanism of the nematocysts of *Pelagia noctiluca*. In: Bolis L, Zadunaisky J, Gilles R (eds) Toxins, drugs, and pollutants in marine animals. Springer, Berling Heidelberg New York, pp 63–76
Salleo A, La Spada G, Alfa M (1983) Blockage of trypsin-induced discharge of nematocysts of *Pelagia noctiluca* by Ca^{2+}. Mol Physiol 3:89–97
Salleo A, La Spada G, Falzea G, Denaro MG (1984a) pH-induced collapse of the capsular wall in isolated nematocysts of *Pelagia noctiluca*. Cell Mol Biol 30:91–94

Salleo A, La Spada G, Falzea G, Denaro MG (1984b) Discharging effect of anions and inhibitory effect of divalent cations on isolated nematocysts of *Pelagia noctiluca*. Mol Physiol 5:25–34

Salleo A, La Spada G, Denaro MG, Falzea G (1986) Effects produced by SCN^- and thioglycolate on isolated nematocysts of *Pelagia noctiluca*. Cell Mol Biol 32:661–666

Salleo A, La Spada G, Denaro MG (1988a) Release of free Ca^{2+} from the nematocysts of *Aiptasia mutabilis* during the discharge. Physiol Zool 61:272–279

Salleo A, La Spada G, Denaro MG, Falzea G (1988b) Dynamics of release of free calcium during the discharge of holotrichous isorhiza of nematocysts of *Pelagia noctiluca*. In: Hessinger DA, Lenhoff HM (eds) The biology of nematocysts. Academic Press, San Diego, pp 551–565

Salleo A, La Spada G, Robson EA (1990) Discharge characteristics of nematocysts isolated from acontia of *Calliactis parasitica*. Mar Biol 104:459–464

Sandberg DM, Kanciruk P, Mariscal RN (1971) Inhibition of nematocyst discharge correlated with feeding in a sea anemone, *Calliactis tricolor* (Leseur). Nature 232: 263–264

Santoro G, Salleo A (1991) The discharge of in situ nematocysts of the acontia of *Aiptasia mutabilis* is a Ca^{2+}-induced response. J Exp Biol 156:173–185

Skaer RJ, Picken LER (1965) The structure of the nematocyst thread and the geometry of discharge in *Corynactis viridis* Allman. Philos Trans R Soc Lond B 250:131–164

Slautterback DB (1967) The cnidoblast-musculoepithelial cell complex in the tentacles of *Hydra*. Z Zellforsch 79:296–318

Smith S, Oshida J, Bode H (1974) Inhibition of nematocyst discharge in hydra fed to repletion. Biol Bull 147:186–202

Stidwill RP, Honegger TG (1989) A single layer of microtubules is part of a complex cytoskeleton in mature nematocytes of hydra. Tissue Cell 21:179–188

Stidwill RP, Honegger TG, Tardent P (1988) Polymerized actin in the apical region of hydra nematocytes. In: Hessinger DA, Lenhoff HM (eds) The biology of nematocysts. Academic Press, San Diego, pp 567–574

Tardent P (1988) History and current state of knowledge concerning discharge of cnidae. In: Hessinger DA, Lenhoff HM (eds) The biology of nematocysts. Academic Press, San Diego, pp 309–332

Tardent P, Holstein T (1982) Morphology and morphodynamics of the stenotele nematocyst of *Hydra attenuata* Pall. (Hydrozoa, Cnidaria). Cell Tissue Res 224:269–290

Tardent P, Zierold K, Weber J, Gerke I (1990a) Metal cations in the nematocysts of cnidaria. Experientia 46:A47

Tardent P, Zierold K, Klug M, Weber J (1990b) X-ray microanalysis of elements present in the matrix of cnidarian nematocysts. Tissue Cell 22:629–643

Thorington GU, Hessinger DA (1988a) Control of discharge: factors affecting discharge of cnidae. In: Hessinger DA, Lenhoff HM (eds) The biology of nematocysts. Academic Press, San Diego, pp 233–253

Thorington GU, Hessinger DA (1988b) Control of cnida discharge: I. Evidence for two classes of chemoreceptor. Biol Bull 174:163–171

Thorington GU, Hessinger DA (1990) Control of cnida discharge: III. Spirocysts are regulated by three classes of chemoreceptors. Biol Bull 178:74–83

Watson GM, Hessinger DA (1988) Localization of a purported chemoreceptor involved in triggering cnida discharge in sea anemones. In: Hessinger DA, Lenhoff HM (eds) The biology of nematocysts. Academic Press, San Diego, pp 255–272

Watson GM, Hessinger DA (1989a) Cnidocytes and adjacent supporting cells from receptor-effector complexes in anemone tentacles. Tissue Cell 21:17–24

Watson GM, Hessinger DA (1989b) Cnidocyte mechanoreceptors are tuned to the movements of swimming prey by chemoreceptors. Science 243:1589–1591

Watson GM, Hessinger DA (1990) Frequency-specificity of vibration-sensitive mechanoreceptors in sea anemone tentacles. Am Zool 30:82A

Watson GM, Mariscal RN (1985) Ultrastructure of nematocyst discharge in catch tentacles of the sea anemone *Haliplanella luciae* (Cnidaria: Anthozoa). Tissue Cell 17:199–213

Weber J (1989) Nematocysts (stinging capsules of Cnidaria) as Donnan-potential-dominated osmotic systems. Eur J Biochem 184:465–476

Weber J (1990) Poly(γ-glutamic acid)s are the major constituents of nematocysts in *Hydra* (Hydrozoa, Cnidaria). J Biol Chem 265:9664–9669

Weber J (1991) A novel kind of polyanions as principal components of cnidarian nematocysts. Comp Biochem Physiol 98A:285–291

Weber J, Klug M, Tardent P (1987a) Some physical and chemical properties of purified nematocysts of *Hydra attenuata* Pall. (Hydrozoa, Cnidaria). Comp Biochem Physiol 88B:855–862

Weber J, Klug M, Tardent P (1987b) Detection of high concentration of Mg and Ca in the nematocysts of various cnidarians. Experientia 43:1022–1025

Weber J, Klug M, Tardent P (1988) Chemistry of hydra nematocysts. In: Hessinger DA, Lenhoff HM (eds) The biology of nematocysts. Academic Press, San Diego, pp 427–444

Westfall JA (1965) Nematocysts of the sea anemone *Metridium*. Am Zool 5:377–393

Westfall JA (1970a) The nematocyte complex in a hydromedusan, *Gonionemus vertens*. Z Zellforsch 110:457–470

Westfall JA (1970b) Ultrastructure of synapses in a primitive coelenterate. J Ultrastruct Res 32:237–246

Westfall JA (1973) Ultrastructural evidence for a granule-containing sensory-motor-interneuron in *Hydra littoralis*. J Ultrastruct Res 42:268–282

Westfall JA, Kinnamon JC (1978) A second sensory-motor-interneuron with neurosecretory granules in *Hydra*. J Neurocytol 7:365–379

Westfall JA, Kinnamon JC (1984) Perioral synaptic connections and their possible role in the feeding behavior of *Hydra*. Tissue Cell 16:355–365

Westfall JA, Yamataka S, Enos PD (1971) Ultrastructural evidence of polarized synapses in the nerve net of *Hydra*. J Cell Biol 51:318–323

Wood RL, Novak PL (1982) The anchoring of nematocysts and nematocytes in the tentacles of hydra. J Ultrastruct Res 81:104–116

Yanagita TM (1943) Discharge of nematocysts. J Fac Sci Tokyo Imp Univ Sec IV 6:97–108

Yanagita TM (1959a) Physiological mechanism of nematocyst responses in sea-anemone. I. Effects of trypsin and thioglycolate upon the isolated nematocysts. Jpn J Zool 12:361–375

Yanagita TM (1959b) Physiological mechanism of nematocyst responses in sea-anemone. II. Effects of electrolyte ions upon the isolated cnidae. J Fac Sci Univ Tokyo Sec IV 8:381–400

Yanagita TM (1960a) Physiological mechanism of nematocyst responses in sea-anemone. III. Excitation and anaesthetization of the nettling response system. Comp Biochem Physiol 1:123–139

Yanagita TM (1960b) Physiological mechanism of nematocyst responses in sea-anemone. IV. Effects of surface-active agents on the cnidae in situ and in isolation. Comp Biochem Physiol 1:140–154

Yanagita TM, Wada T (1953) Discharge-inducing concentrations of acids and bases for the nematocysts of sea-anemone. Nat Sci Rep Ochanomizu Univ 4:112–118

Yanagita TM, Wada T (1954) Effects of trypsin and thioglycolate upon the nematocysts of the sea anemone. Nature (Lond) 23:171

Yu S-M, Westfall JA, Dunne JF (1985) Light and electron microscopic localization of a monoclonal antibody in neurons in situ in the head region of *Hydra*. J Morphol 184:183–193

Zierold K, Gerke I, Schmitz M (1989) X-ray microanalysis of fast exocytotic processes. In: Zierold K, Hagler HK (eds) Electron microprobe analysis: applications in biology and medicine. Springer, Berlin Heidelberg New York, pp 281–292

Chapter 3

Physiological Adaptations in Decapodan Crustaceans for Life in Fresh Water

Michele Wheatly

Contents

Department of Zoology, University of Florida, 233 Bartram Hall, Gainesville, FL 32611, USA

Advances in Comparative and Environmental Physiology, Vol. 15

1 Introduction

In recent years, there have been several excellent review articles on aspects of decapod crustacean physiology including ventilation and circulation (Taylor 1982; Cameron and Mangum 1983; McMahon and Wilkens 1983; McMahon and Burggren 1988), acid-base balance (Truchot 1983; Cameron 1986), gas transport (McMahon 1981; Mangum 1983), and osmoregulation (Mantel and Farmer 1983). These have focused predominantly on the marine decapods that constitute the majority (90%) of crustacean species. While research has steadily continued on freshwater (FW) decapods such as the crayfish, this information is typically "lost" among the wealth of information on marine species. This is regrettable because FW species exhibit some of the most sophisticated physiological mechanisms among crustaceans. To name but a few, they have well-developed branchial ion uptake mechanisms, a kidney with the unique ability to produce dilute urine, and adaptations for molting and postmolt calcification in an inhospitable environment. Furthermore, the physicochemical properties of FW dictate that environmental challenges such as hypoxia, hyperoxia, hypercapnia, and aerial exposure, as well as man-made problems such as acidification, are experienced more routinely by FW as opposed to marine species. In summary, therefore, it would appear that FW decapods deserve separate recognition.

Approximately 10% of decapodan species are indigenous to FW environments, including species of shrimp, crayfish, and crabs. In addition, there are many marine species that can tolerate FW to varying degrees, notably *Eriocheir*, *Rithropanopeus*, and *Callinectes*; however, they must return to sea water (SW) to breed. For the present purposes, this review article is restricted to physiological adaptations in aquatic species which complete their entire life cycle in FW. There are also many FW species which routinely encounter aerial exposure due to the nature of inland FW environments (see below), and some decision has to be made regarding the extent to which terrestrial species of FW origin will be covered in this article. For this, I decided to follow the logic presented in the preface to Burggren and McMahon (1988a) in their consideration of what constituted a "land crab" and so have restricted this review to species which spend a significant amount of their time and activity in FW. To begin with, I will introduce the cast of characters and comment briefly on their biogeography and evolution.

2 Biogeography and Evolution of Freshwater Decapods

There have been numerous invasions of FW by decapods, but with the exception of the reptant (nonswimming) crayfish (infraorder Astacidea, families Potamobiidae and Parastacidae) and the true FW crabs (infraorder Brachyura, section Brachyrhyncha, superfamily Potamoidea), these have not been particularly successful. Among the natant (swimming) shrimp (infraorder Caridea), there are some families that include FW members such as the Atyidae (e.g., *Atya*) and the Palaemonidae (e.g., *Palaemonetes*, *Macrobrachium*). However, their distribution points to a very recent immigration into FW. Furthermore, they have a poorly preserved fossil record (Jurassic-Recent; Schram 1982) due to their swimming habit, which demands a less heavily sclerotized cuticle.

Of the two major groups of FW decapods, the crayfish are geologically older (Abele 1982). Moreover, they do not generally coexist with the FW crabs. FW crayfish are restricted to temperate areas, while FW crabs occur in tropical parts, and there appears to be little overlap in their distribution. Ortmann (1902), in an excellent article which readers are urged to consult, incorporated the fossil record, phylogenetic relationships, paleogeography, and present day distribution into an analysis of the geographical distribution of FW decapods. Neither FW crayfish nor FW crabs possess any exceptional means of dispersal, since they are restricted to FW and cannot leave for extended periods to migrate overland. Additionally, the egg and larval stages do not permit passive transport. Crayfish also exhibit direct development and some parental care which further limit dispersal.

The reptant infraorders have good fossil records due to their more heavily mineralized skeletons. The Astacidea can be traced back to the Late Permian. In the Lower Cretaceous period, the ancestors of Potamobiidae and Parastacidae lived in Sino-Australia. When the separation of E. Asia and Australia occurred in the Upper Cretaceous, the two became differentiated. Parastacidae subsequently became established by the Upper Tertiary in relatively isolated areas of Australia, New Zealand, and South America, where the family is represented by such species as *Cherops*, *Astacopsis*, *Engaeus*, *Parastacus*, and *Astacoides*, among others. Potamobiidae spread from East Asia east to America, and west to West Asia and Europe, where it is represented by the two genera *Potamobius* and *Cambarus*.

The ancestral form of Potamonidae was probably an estuarine xanthid crab which colonised FW in Asia/India/Africa in the Late Cretaceous/

Early Tertiary. While the African stock remained essentially unchanged, the Indian stock spread into America, Australia, and the Mediterranean. The family Potamonidae is represented by four subfamilies: (i) Potamoninae, including *Parathelphusa* and *Potamonautes* in India, Asia, and Africa, (ii) Deckeniinae, including a single genus found in East Africa and the Seychelles, (iii) Potamocarcininae, including *Pseudothelphusa* in America, and (iv) Trichodactylinae including *Trichodactylus* in South America. The arrival of FW crabs into an area occupied by crayfish (such as India and SE Asia in the Upper Cretaceous) was generally followed by the extermination of crayfish in that region, possibly because crabs are generally more active. In more temperate regions of China, Australia, and Mexico, the advance of crabs was probably halted by climatic barriers.

Other studies of FW decapod zoogeography have been geographically more restrictive, such as the one by Bishop (1967) of Australia. He identifies two major families of Australian FW decapods, the Parastacidae and Hymenosomatidae, which contains a single species. The former arrived as a result of continental drift in the Early Mesozoic. The remaining three families that are represented in Australia (Atyidae, Palaemonidae, and Potamonidae) are all circumtropical in origin and arrived from the north via a series of island hopping or by land bridge crossing. This was somewhat easier for the first two because they have planktonic dispersive larval stages in their life history, whereas potamonids do not. SE Australia appears to be the site of the major FW decapod diversification in terms of genera and species found.

As will become apparent during the course of this review, the physiological adaptations for life in FW are variable, which would confirm that evolution into FW was polyphyletic.

3 Physicochemical Challenges of Living in Fresh Water

While many physicochemical characteristics differ in FW and SW, probably the most important distinction is that FW habitats are greater in variety and complexity than SW, making it hard to define "average" conditions (Pennak 1989). This probably arises from the fact that FW habitats, which cover less than 2% of the earth's surface (compared with 70% under SW), are geographically fragmented, whereas the marine environment is continuous. In addition, the former can include both running or standing water, which will affect the degree of mixing.

The most obvious difference relates to total inorganic ion content (Greenaway 1979), which, expressed as osmolality, is as low as $4\,\mathrm{mOsm\,kg^{-1}}$ in FW as opposed to around $1000\,\mathrm{mOsm\,kg^{-1}}$ in SW. Not only does this reflect reduced ion concentrations (for example, NaCl is around $0.3\,\mathrm{mmol\,l^{-1}}$

in FW as opposed to 500 mmol l^{-1} in SW) but also changes in the relative proportions of ions. For example in FW, $CaCO_3$ may constitute more than 50% of the total dissolved solids (TDS), whereas there is less than 1% $CaCO_3$ in SW, while 80% is attributable instead to NaCl (compared with 5% in FW). Moreover, the precise levels are very variable in FW, depending upon run-off. Generally speaking, FW encompasses TDS ranging from 10 to 1000 mg l^{-1}, and many permanent populations can tolerate this 10-fold range. Running water is especially notable for the annual variations in dissolved materials (±80%), depending upon the local geochemistry and seasonal fluctuations in discharge.

Respiratory gases may also vary considerably, both spatially and temporally, in FW environments, especially in smaller, stagnant, unmixed water bodies with plant and animal populations (Truchot 1987). Here, diurnal changes in O_2 concentration may result from respiratory and photosynthetic processes, leading to hypoxia during the night and hyperoxia by day, with accompanying changes in CO_2 levels. Very high CO_2 values can also develop during the anaerobic fermentation of large amounts of decaying organic matter in environments where light penetration and hence photosynthesis are prevented.

Another variable property of FW is pH, which can vary from 4.4 to 8.6 depending on the vagaries of photosynthesis, light, current, respiratory processes, biota, and circulation. Small inland bodies of water are also more susceptible to drying up, depending upon climatic conditions, so that FW species are more likely to experience aerial exposure.

Due to the relatively small volume of inland waters and their reduced buffering capacity, as well as their close proximity to human habitation, pollution by both domestic and industrial waste has far greater ramifications in FW than SW. Resulting waters have low O_2, high organic matter, and extreme pH.

Although the overall picture of FW environments is one of high variability, I should add that subterranean waters are somewhat of an exception to this, possessing a greater degree of constancy.

4 Osmoregulation and Ionoregulation

Obviously, the most pronounced difference between FW and SW is in ionic composition, and so osmo- and ionoregulatory adaptations will be considered first. While the majority of marine decapods are isosmotic to SW, FW species have to maintain body fluid concentrations above external levels, which results in diffusive water gain and ion loss. In order to offset these passive tendencies, decapods generally exhibit reductions in permeability and mechanisms to take up or retain ions actively. The ion balance has been studied extensively in FW crayfish (see review by Wheatly 1993) and less well studied in FW shrimp (Parry 1957) and FW crabs (Shaw 1959a; Lutz 1969; Harris and Micallef 1971; Greenaway and MacMillen 1978). FW species commonly attempt to reduce the osmotic and ionic gradients they maintain with the external medium, in order to minimize the energetic cost of ionoregulation. Thus, extracellular (EC) osmolality is typically 250–500 mOsmol kg^{-1} in FW crayfish and shrimp; values in FW crabs are higher at 500–600 mOsmol kg^{-1}, perhaps indicating a more recent invasion of FW (Table 1A, Mantel and Farmer 1983). Both are approximately one-half of the value found in isosmotic marine species.

The difference in EC osmolality between FW crayfish and FW crabs is similarly reflected in levels of individual ions, at least with respect to Na^+, K^+, and Cl^-, which are all lower in FW crayfish (see Table 11, Greenaway 1979); calcium levels, which are virtually identical in both FW groups (around 12 mmol l^{-1}), are very similar to levels reported in SW species, suggesting that Ca is regulated at an absolute value in the majority of Crustacea.

4.1 Passive Permeability

Hyperosmo/ionoregulation is accomplished by a combination of mechanisms to reduce passive (primarily branchial) permeability to ions and water (thereby slowing ion efflux and water influx) together with active branchial mechanisms to take up ions and renal mechanisms to eliminate accumulated water. For some decapod crustaceans, such as the crayfish, data are available on each of these mechanisms; however, the research is incomplete on many of the other FW decapods.

The ionic permeability in FW decapods is generally lower than in marine species. The sodium permeability for marine decapods ranges from 0.05 to 0.30 l kg^{-1} h^{-1} (Shaw 1961; Cornell 1976), whereas in FW decapods it is an order of magnitude lower (0.001–0.005 l kg^{-1} h^{-1}; Shaw 1959a; Bryan 1960). Specific values for FW decapods range from 0.1 in the shrimp *Palaemonetes antennarius* (Parry and Potts 1965) to 0.008 in the FW crab *Potamon edulis* (Harris 1975) and 0.005 in the FW crayfish *Austropotamobius pallipes* (Sutcliffe 1975), suggesting that a reduction in ion permeability may reflect evolutionary history in FW.

Permeability coefficients for water (in l $mOsmol^{-1}$ h^{-1}) are lower in FW decapods (0.0004–0.009) than in marine forms (0.1–0.5), but this time by 2 orders of magnitude (Kirschner 1979). Specific values indicate that osmotic permeability is lower in *Potamon niloticus* (Shaw 1959a) than in *Astacus fluviatilis* (Bryan 1960). This may relate to the efficiency of renal elimination of the water load in the two species. Since FW crabs are unable to produce a dilute urine (see Sect. 4.3 below), it behooves them to restrict water uptake; because FW crayfish can produce a hypotonic urine, they can excrete water without exacerbating their ion loss problem, and so their urine flow rate is higher.

4.2 Branchial Uptake Mechanisms

Branchial ion uptake has been well documented especially in FW crayfish. Early work by Bryan (1960) and Shaw (1960a,b) indicated that the crayfish actively absorbs Na^+ and Cl^- from the medium via ion exchange pumps located on the gills. The counterions for these uptake mechanisms are believed to be H^+ or NH_4^+ in the case of Na^+, and HCO_3^- or OH^- for Cl^- (Ehrenfeld 1974), establishing an important link between ion uptake and acid-base regulation which will be explored further below (Sect. 5.1.3). While the Na^+ and Cl^- uptake mechanisms are independent, they appear to be linked (Shaw 1964); both transporters are sensitive to acetazolamide, suggesting that the enzyme carbonic anhydrase (CA) provides counterions for both uptake mechanisms from respiratory CO_2.

Radioisotope methodology has been used to resolve the net branchial Na^+ and Cl^- uptake (in each case around $+50\,\mu mol\,kg^{-1}\,h^{-1}$) into unidirectional components in crayfish (Wood and Rogano 1986; Wheatly 1989). In the case of Na^+, a unidirectional uptake of +300 is countered by a unidirectional efflux (presumed passive) of $-250\,\mu mol\,kg^{-1}\,h^{-1}$. The Cl^- unidirectional fluxes are approximately double these values (influx +630, efflux $-580\,\mu mol\,kg^{-1}\,h^{-1}$). In the case of each ion, the small net branchial uptake counters the urinary net efflux (see below), thus achieving ion balance. Parry and Potts (1965) made similar measurements in the FW shrimp *Palaemonetes antennarius*, except for whole-body fluxes which have not been resolved into branchial versus renal components. They reported equivalent, whole animal, unidirectional, Na^+ influx and efflux of $3300\,\mu mol\,kg^{-1}\,h^{-1}$, which are both an order of magnitude greater than in crayfish. This may be explained by the fact that shrimp have more recently invaded FW and thus have flux rates characteristic of marine species which tend to be quantitatively greater due to extensive exchange diffusion. The efflux component will be largely attributable to urinary losses, and this will be elaborated below.

Ultrastructural evidence that ion transporting cells are located on the gills in crayfish was provided by Fisher (1972). Chloride cells are believed to be the site of ion transport, and these are characterized by a columnar shape, apical pit, high density of mitochondria, and extensive infoldings of the basal membrane. In addition, they possess high levels of activity of the two transport marker enzymes: (Na^+/K^+)-ATPase and CA. In crayfish, chloride cells and transport enzyme activities are uniformly distributed among the gill pairs (Dickson and Dillaman 1985; Wheatly and Henry 1987), again confirming that all gills function in ion and acid-base regulation, which may relate to the increased demand for ion uptake from an ionically limited environment. The subcellular location of the ion transporters in hyperosmoregulating species was reviewed by Kirschner (1979), and the role of CA in crustaceans was elaborated by Henry (1984). The amiloride-sensitive Na^+/acidic equivalent and SCN-sensitive Cl^-/basic equivalent ion exchangers are located on the apical surface of the branchial cell; once inside the cell, Na^+ is moved into the hemolymph via the classical ouabain-sensitive basolateral (Na^+/K^+)-ATPase, and Cl^- follows by some unknown mechanism which is presumed to be passive and may be favored by the electrochemical gradient. In marine crabs, the transport enzyme activity is concentrated in the posterior two or three gill pairs (Henry and Cameron 1982) where the chloride cells are located (Copeland and Fitzjarrell 1968). The Na^+ and Cl^- uptake mechanisms in crayfish are saturable, displaying K_m (external concentration at half-saturation) values of 0.25 and 0.05 m*M*, respectively (Shaw 1959b, 1960b). Shaw (1959a) also measured the K_m for Na^+ uptake in the FW crab *Potamon niloticus* and found it to be virtually identical to that of the crayfish.

Most of the research on ion regulation has focused on the major hemolymph ions Na^+ and Cl^-, with a little attention being paid to some of the less prevalent EC ions. Calcium is very important in the context of molting physiology, and variation throughout the molting cycle will be addressed separately below (Sect. 6). Unlike Na^+ and Cl^-, intermolt crayfish are in negative Ca^{2+} balance, losing Ca^{2+} into the medium at a rate of approximately $-40\,\mu mol\,kg^{-1}\,h^{-1}$, primarily at the gills (Greenaway 1972). Since the carapace constitutes a sizeable reservoir of Ca^{2+}, this loss is negligible and may even represent an energetic saving in that metabolic energy is not required for active uptake. As for other electrolytes, K^+ absorption has been demonstrated in the crayfish by Shaw (1960b), and Wheatly (1989) recently showed branchial Mg^{2+} uptake also in crayfish; however, the cellular mechanisms for the uptake of these ions have not been identified. Regulation of the anions $SO_4{}^{2-}$ and $PO_4{}^{2-}$ has not been investigated.

4.3 Urinary Mechanisms

The general structure of each of the paired crustacean antennal glands is an end-sac or coelomosac, which encloses the remnants of the coelom, an excretory canal, which is subdivided into a labyrinth and tubule, and an exit duct which expands into a large bladder and opens onto the third or antennal segment in decapods (Riegel 1972; Mantel and Farmer 1983). In FW crayfish, the tubule is differentiated into a proximal and distal section, and the latter is the site of active NaCl reabsorption (Peterson and Loizzi 1974), enabling the crayfish to produce dilute urine. Additional ion reabsorption occurs in the bladder (Kamemoto et al. 1962).

The rate of urine flow is correlated with osmotic permeability (see Sect. 4.1). Thus, in *Potamon edulis* and *P. niloticus* it is only 0.25 ml kg^{-1} h^{-1} (Shaw 1959a; Harris 1975), whereas in the FW crayfish it is 2–4 ml kg^{-1} h^{-1} (Bryan 1960; Riegel 1963; Greenaway 1970; Tyler-Jones and Taylor 1986; Wheatly and Toop 1989). A study on the FW shrimp *Palaemonetes antennarius* (Parry 1957) indicates that the urine flow rate is as high as 20 ml kg^{-1} h^{-1}, although the corresponding water permeabilities are not available.

Freshwater decapods vary in their ability to dilute the urine, which is desirable for any hyperregulator in that it rids the body of water without compounding the ion loss problem. Shaw (1959a) found that the FW crab *Potamon niloticus* was capable of producing a urine osmolality of 450 compared with 500 mOsm kg^{-1} in the hemolymph, and Harris and Micallef (1971) reported the production of isosmotic urine in *P. edulis*. Freshwater shrimp appear more variable in their urinary function, with some only reabsorbing 0–10% of filtered electrolytes (*Palaemon longirostris*, *P. antennarius* and *P. varians*; Parry 1957), while others are capable of 80–90% reabsorption (*Syncaris pacifica*, Born 1968). The crayfish appear to exhibit the most effective (95%) ion reabsorption of all (*Austropotamobius pallipes*, Riegel 1968; *Pacifastacus leniusculus*, Kerley and Pritchard 1967).

The FW crabs and shrimp, which produce essentially isosmotic urine, have urine: hemolymph ratios for Na^+ of virtually 1 (Parry 1957; Harris and Micallef 1971). A detailed study of postfiltrational electrolyte reprocessing has been undertaken in the crayfish (Wheatly and Toop 1989) which indicates that, under control conditions, the ions most effectively reabsorbed are phosphate and sulfate (97%), closely followed by Na^+, Ca^{2+}, and Cl^- (at 95%). K^+, Mg^{2+}, and HCO_3^- were less effectively reabsorbed (85, 72, and 65%, respectively).

The mechanism for Na^+ reabsorption at the renal tubule cell involves (Na^+/K^+)-ATPase (Peterson and Loizzi 1974) and CA (Wheatly and Henry 1987). Recently, Ca^{2+}-ATPase activity was also detected at the antennal gland (Wheatly 1990). Thus, the transport of electrolytes probably involves similar mechanisms to those identified at the branchial epithelium.

The relative involvement of the antennal gland versus gills in ion regulation has been the focus of debate. Cellular marker enzyme specific activities [(Na^+/K^+)-ATPase and CA] were both higher in the antennal gland than at the gill in the crayfish (Wheatly and Henry 1987), whereas in marine species they are typically lower (Henry and Cameron 1982), commensurate with the production of isosmotic urine. This was further confirmed in the study by Wheatly and Toop (1989), in which the calculated renal reabsorption rates of Na^+ and Cl^- in crayfish were threefold greater than the branchial unidirectional influx rates, suggesting that the kidney plays a greater role in ion uptake than the gill, at least for Na^+ and Cl^-. This also suggests that the reabsorption of ions from the urine is more economical than replacing lost ions from the medium at the gill, where the electrochemical gradient is larger.

The net renal efflux of Na^+ and Cl^- is approximately $-80\,\mu mol\,kg^{-1}\,h^{-1}$, which is offset by a net branchial uptake of $+50\,\mu mol\,kg^{-1}\,h^{-1}$ (Wheatly 1989), achieving a whole animal ion balance. The net renal effluxes of Na^+ and Cl^- constitute 32 and 14%, respectively, of the branchial unidirectional efflux. Obviously, in FW crabs producing isosmotic urine, the total urinary NaCl loss is an order of magnitude greater than in the FW crayfish (Greenaway 1979); however, due to large branchial unidirectional fluxes, it still only constitutes around 10% of the total ion loss.

Some data are also available for Ca^{2+} regulation. In the intermolt crayfish, the renal Ca^{2+} efflux is $-7\,\mu mol\,kg^{-1}\,h^{-1}$ (Wheatly and Toop 1989), which constitutes about 20% of the whole-body Ca^{2+} loss (Greenaway 1972). However, in this case, since there is no active uptake at the gill during intermolt, the antennal gland is solely responsible for Ca^{2+} reabsorption, which continues at a rate of $+70\,\mu mol\,kg^{-1}\,h^{-1}$. Not surprisingly, the Ca^{2+}-ATPase activity in the antennal gland of intermolt crayfish is double that measured at the gill (Wheatly 1990).

4.4 Tissue Ion Regulation

Intracellular (IC) inorganic electrolyte concentrations have been estimated for the FW crayfish. The abdominal muscle of *Pacifastacus leniusculus* has the following ion concentrations (in m*M* kg^{-1} IC H_2O: Na^+ 22; K^+ 105; and Cl^- 12; M.G. Wheatly and E.C. Vevera, unpubl. observ.). These are appreciably lower than corresponding values for marine decapods (Wheatly 1985a; Na^+ 89; K^+ 169; and Cl^- 123), presumably due to lowering of the EC osmolality in FW-adapted species. That study also compared the inorganic electrolyte levels in a variety of different tissues. Levels of Na^+ and Cl^- were greatest in nerves, intermediate in the heart, and least in muscle, agreeing with studies on other species. There also appeared to be variation between different muscle types, with the slow claw closer muscle having higher Na^+ and Cl^- levels than the fast abdominal flexor muscle, again agreeing with work on other species. As with whole animal epithelial exchange, there is believed to be a strong connection between the ion and acid-base regulation at the cell membrane via electroneutral exchanges (i.e., Na^+/acidic equivalent; Cl^-/basic equivalent). These will be outlined below (Sect. 5.2.3) in the context of the acid-base balance.

5 Acid-Base Balance

The acid-base balance is essential in both the EC compartment, for the proper function of respiratory proteins, and in the IC compartment, for the preservation of enzyme-mediated cell metabolism. In all animals, it involves interactions between a number of physiological and biochemical processes including respiratory gas exchange, intermediary metabolism, and, in the case of aquatic species, ion transport. Among the crustaceans, the EC acid-base balance has been reviewed by Truchot (1983), Cameron (1986), and in the extensive monograph by Truchot (1987). None of these reviews includes a significant treatment of the IC pH, since most of this research has only been recently conducted, although some information is included in a recent review by Henry and Wheatly (1992). Due to the exhaustive nature of this subject, coverage in this review article will be limited to highlighting the mechanisms of EC and IC acid-base regulation in FW species, and their relative importance in the whole-animal acid-base homeostasis. Little attention will be given to the theoretical treatment of the acid-base balance, which is covered extensively elsewhere (e.g., Truchot 1987; Cameron 1989a).

5.1 Extracellular Acid-Base Balance

5.1.1 Normal Profile

The EC acid-base profile in FW crustaceans such as the crayfish is fairly typical of values reported for marine species at comparable temperatures with a pH between 7.7 and 8.0, PCO_2 between 2 and 4 torr and HCO_3^- concentration between 4 and 10 m*M* (Truchot 1987). In a FW species, the EC acid-base balance is highly dependent upon the concentration of the external Na^+ and Cl^-. As outlined above (Sect. 4.2), the Na^+ and Cl^- uptake mechanisms are via branchial electroneutral ion exchangers (Na^+/H^+; Cl^-/HCO_3^-), which establish a link between ion and acid-base regulation. Thus, changes in the ambient Na^+ or Cl^- concentration below the range for Michaelis-Menton saturation kinetics will perturb the acid-base balance by inhibiting the electroneutral exchanges. In the crayfish *Astacus leptodactylus* (Dejours and Armand 1982; Dejours et al. 1982), a decrease in the external Cl^- below 1 m*M* induced a metabolic alkalosis which correlated well with the kinetics of the Cl^- uptake mechanism ($K_m = 100\,\mu M$; Shaw 1960a). In another study (Short and Haswell 1979), exposure to Na^+-free water led to acidosis. It is highly probable that other external ions may also affect the EC acid-base balance, such as Ca^{2+} which can affect Na^+ fluxes, but these have not been tested so far.

5.1.2 Extracellular Acid-Base Regulation

As mentioned in the introductory remarks, FW is variable in its physico-chemical characteristics, and FW decapods will be required to regulate their EC acid-base balance in response to such parameters as levels of respiratory gases, external acid-base balance, and temperature. Typically, the exposure of crayfish to lowered ambient O_2 (hypoxia) is associated with a respiratory alkalosis (Sinha and Dejours 1980; Wheatly and Taylor 1981; Wilkes and McMahon 1982a) that is compensated to varying degrees over time by a reduction in the metabolic HCO_3^- concentration. Conversely, environmental hyperoxia (Gaillard and Malan 1983; Wheatly 1989) is characterized by a hemolymph acidosis compensated by the accumulation of metabolic HCO_3^-. Both responses are governed by the fact that water oxygenation regulates ventilation in water-breathers. Freshwater crayfish also show a response to external PCO_2. Environmental hypercapnia was accompanied by a hemolymph acidosis and attendant increase in metabolic HCO_3^- concentration in crayfish (Dejours and Armand 1982; Gaillard and Malan 1983).

The EC acid-base balance is also dependent upon the acid-base status of the water, such that a decrease in carbonate alkalinity causes a hypercapnic acidosis due to a reduced CO_2 capacitance coefficient and gill CO_2 conductance (Dejours and Armand 1982). This can be compensated metabolically, as long as NaCl is not limited in the water (Burtin et al. 1986). The effects of reducing the water carbonate alkalinity are probably not linked to the weak acidification that occurs. However, specific pH effects (metabolic acidosis) do occur in response to extreme acid exposure (see Sect. 9.2). This results from acid entry and ionic disturbance and seems to be ameliorated somewhat in soft as opposed to hard water. In the FW crayfish that have been examined, there appears to be no compensatory response (Morgan and McMahon 1982; Wood and Rogano 1986).

In the crayfish, the EC pH varies inversely with temperature according to the temperature-pH slope ($\Delta pH/\Delta t$) of -0.0165 pH units $°C^{-1}$ (Dejours and Armand 1983; Gaillard and Malan 1985). The increase in pH results from an increased HCO_3^- at an approximately constant PCO_2 accomplished via ion transport mechanisms. The PCO_2 changes only slightly due to temperature effects on the solubility and pK of the CO_2/HCO_3^- system.

Due to differences in the gas characteristics of water and air, air-breathing crustaceans (and water-breathers exposed in air) tend to have a higher PCO_2 and HCO_3^- and correspondingly a lowered pH. However, PCO_2 values are not as high as in air-breathing vertebrates. This has led some authors (e.g., Dejours and Truchot 1988) to hypothesize that crustaceans in air hyperventilate in order to oxygenate sufficiently the respiratory pigment hemocyanin (HCy). This may be due to inefficient perfusion or gill lamellae collapse which results in diffusion-limited gas exchange. In the FW land crab, *Holthuisana*, in water at 25 °C, Greenaway et al. (1983a) reported the following values: pH 7.35; PCO_2 6.4 torr; and HCO_3^- 9.7 m*M*. When these crabs were breathing in air (Greenaway et al. 1983b), the values were: pH 7.36; PCO_2 10.5 torr; and HCO_3^- 13 m*M*. Values have also been reported for the Trinidadian FW land crab *Pseudothelphusa garmani*: PCO_2 6 torr and total CO_2 8.4 m*M* (Innes and Taylor 1986). When water-breathing FW crustaceans such as crayfish were exposed to air (Wheatly and Taylor 1981), the pH dropped from 7.9 initially to 7.5, associated with an increase in PCO_2 from 3 to 9 torr and an increase in lactate from 1 to 8 mM. Over 24 h, this mixed respiratory and metabolic acidosis was completely compensated by an accumulation of HCO_3^- from 7 to 14 mM and recovery of the lactate levels.

The majority of air-breathers regulate their EC acid-base balance by respiratory control of PCO_2 (Truchot 1987). While air-breathers may respond to a lack of O_2 by increased ventilation, causing a hypocapnic alkalosis, the response is generally less sensitive than in water-breathers. Lung ventilation is typically dependent on internal CO_2 and pH levels, and increased inspired PCO_2 leads to respiratory acidosis which is alleviated by increased ventilation. There may also be some progressive increase in the metabolic HCO_3^- value. However, the only studies on air-breathing crustaceans have been on marine species.

There are other EC acid-base disturbances which do not depend upon the physicochemical properties or nature of the respiratory medium and instead are endogenously generated. Two examples are exercise and molting. The effects of exercise (running and tail flipping) have been evaluated in the FW crayfish *Procambarus* by McMahon (1981) at 10 °C. After exercise, the pH dropped from 8 to 7.6 accompanied by an increase in PCO_2 from 2.3 to 3.1 torr (accounting for 70% of the pH disturbance) and an increase in lactate from 0.1 to 1.6 mM (30%). The increase in PCO_2 may result from an increased CO_2 production, inability of the gills to excrete CO_2 rapidly enough, and disturbance in the equilibrium between H^+ and HCO_3^-. Because exercise acidosis is associated with hyperventilation, it generally results in a loss of total CO_2. However, in the crayfish study, the total CO_2 level stayed approximately constant. There is a distinct possibility in FW crustaceans that the massive hyperventilation during exercise exacerbates the branchial ion loss. It is conceivable that some limitation of branchial exchange occurs to reduce this passive ion loss. Unfortunately, it is difficult to draw generalizations for FW crustaceans since few exercise studies have been undertaken. Usually, the acid-base balance is reestablished rapidly on recovery in water.

As outlined below (Sect. 6) the molting cycle in crustaceans is characterized by a resorption of $CaCO_3$ from the carapace during premolt and deposition during postmolt. Thus, in the FW crayfish (Dejours and Beekenkamp 1978), the hemolymph total CO_2 and pH rise prior to ecdysis and decrease immediately thereafter. Since there are no accompanying changes in PCO_2, the acid-base disturbance is assumed to be primarily metabolic.

5.1.3 Mechanisms of Extracellular Acid-Base Regulation

In aquatic crustaceans, there are two primary mechanisms of EC acid-base regulation which are (i) buffering of protons by organic/inorganic buffers and (ii) transport of acid/base equivalents from the hemolymph to either the external water or the IC fluid compartment. The most important EC buffer is HCy, which can be titrated at constant PCO_2. However, there is no evidence that the total EC protein is regulated for acid-base purposes.

Ion regulation with the external environment is the primary mechanism of acid-base regulation in water-breathers, and this occurs at two primary sites: the gill, which is a bidirectional exchanger, and the antennal gland, which can alter the rate at which electrolytes are lost (effectively efflux). The kidney does exhibit unidirectional influx, but in this case from the luminal fluid. In the FW crayfish, all of the gill pairs (13–18 pairs) appear to be involved in ion transport and thus acid-base regulation. Also, the antennal gland plays a more significant role in acid-base balance since it has a high ion transporting capability.

5.1.3.1 Branchial

As outlined above (Sect. 4.2) indirect evidence for the coupling of ion and acid-base exchange in FW crayfish came from early work by Shaw (1959b, 1960a,b) who demonstrated that H^+ excretion varied as a function of external Na^+ concentration and that a high external H^+ level inhibited Na^+ uptake. He also showed that an elevated external NH_4^+ concentration inhibited Na^+ uptake. He also showed that an elevated external NH_4^+ concentration inhibited Na^+ uptake. Ehrenfeld (1974) and Kirschner et al. (1973) demonstrated a correlation between Na^+ influx and H^+ excretion which was amiloride-sensitive, but only in the ion-depleted crayfish. Ehrenfeld (1974) also demonstrated a correlation between Cl^- uptake and excretion of titratable base, presumably HCO_3^-, which was also acetazolamide-sensitive. Thus, indirect evidence suggests that H^+ and NH_4^+ (acidic equivalents) may act as counterions for Na^+ uptake, and HCO_3^- (or other basic equivalents) for Cl^- uptake. The counterions originate from respiratory CO_2 as outlined above (Sect. 4.2), implicating the involvement of the enzyme CA. Although this original work laid the foundation for future research, the specifics are questionable, since most of the studies used ionically depleted animals and failed to separate the renal and branchial components of the whole-animal response.

What was required to confirm the connection between ion and acid-base regulation was direct evidence that these electroneutral ion exchangers are dynamically adjusted to achieve acid-base homeostasis; this has been demonstrated in two recent studies using FW crayfish. In the first one, Wood and Rogano (1986) found that exposure of the crayfish *Orconectes propinquus* for 5 days to acid decarbonated soft water caused an EC metabolic acidosis accompanied by moderate depressions in the EC Na^+ and Cl^- contents due to losses to the water. The acidosis was attributed to the whole-body uptake of acidic equivalents from the environmental water. Unidirectional radiotracer flux analysis demonstrated that a negative Na^+

and Cl^- balance resulted from partial inhibition of the influx components; unidirectional effluxes were not affected. These authors concluded that the inhibition in Na^+ influx and increase in H^+ net flux could have resulted from H^+ versus Na^+ competition for a common carrier. The inhibition in Cl^- influx may have reflected conformational changes in the carrier and/or reduction in the internal HCO_3^- level impeding the Cl^-/HCO_3^- exchange. While the electroneutral ion exchangers accounted for part of the Na^+ and Cl^- uptake, a large portion of the unidirectional fluxes of both ions appeared to be via exchange diffusion.

In the second study, Wheatly (1989) exposed the crayfish *Pacifastacus leniusculus* to environmental hyperoxia for 72 h. In this study, the antennal glands were cannulated, enabling the branchial fluxes to be studied in isolation. An initial EC respiratory acidosis was compensated within 48 h by the accumulation of metabolic HCO_3^- accompanied by branchial net H^+ output. At the same time, the EC Cl^- level was reduced due to net Cl^- efflux. The latter was attributed to an initial increase in efflux followed by a reduction in influx. Meanwhile, the crayfish remained in Na^+ balance, although there were parallel reductions in both the influx and efflux components, suggesting a decrease in exchange diffusion. In a study on a marine land crab (*Cardisoma*; Wood and Boutilier 1985), the net H^+ flux could be explained entirely in terms of the difference between the measured strong cation and anion fluxes. This was not true for the crayfish study, suggesting that other unmeasured ions were involved. Of the two major hemolymph electrolytes, the Cl^- net and unidirectional fluxes had the best correlation coefficients with the net H^+ flux. There was also a good correlation between the net H^+ flux and differences between the Na^+ and Cl^- unidirectional effluxes and influxes. As in the earlier study, much of the unidirectional flux of both ions was via exchange diffusion, suggesting that only a small part may represent exchange for acidic/basic equivalents.

Burtin and Massabuau (1988) recently demonstrated that in FW crayfish the CO_2 exchange system can participate in pH regulation (much as in air-breathers) in situations in which the ion balance is impaired. In their study, they observed that hypercapnic exposure induced an acidosis which was compensated metabolically. However, when combined with exposure to low environmental NaCl, it led to ventilatory compensation. Thus, there is no hard-and-fast rule regarding routes for acid-base regulation in air- versus water-breathers.

5.1.3.2 Renal

Assuming that the acid-base and electrolyte exchange are linked at the renal tubule cells as at the branchial epithelium and knowing that the antennal gland in the crayfish has a high ion transporting capability, then this organ could potentially be involved in acid-base homeostasis. This was investigated during the hyperoxic episode outlined above in a study by Wheatly and

Toop (1989). That study and a historic measurement by Maluf (1941) indicate that crayfish produce a urine that is acid relative to the hemolymph with 65–95% recovery of filtered electrolytes. Marine species produce a relatively alkaline isosmotic urine (Cameron and Batterton 1978; Wheatly 1985b). Urine pH appears to be primarily determined by nontitratable (ammonia) components in FW species and by titratable components in marine species. During hyperoxic exposure, the crayfish urine became increasingly acidified, mirroring the branchial response. Quantitatively, however, the magnitude of the renal response was only 10% of the net branchial excretion, suggesting that it plays a relatively minor role in the whole-animal acid-base response. When the roles of the two exchange epithelia in ion transport were compared, the net effluxes of Na^+ and Cl^- tended to counter each other during hyperoxia. Although the electrolyte excretion increased at this time due to the increased urinary flow, the renal reabsorption rates typically doubled, suggesting that the glomerular electrolyte filtration and tubular reabsorption were balanced. Renal HCO_3^- reabsorption was elevated fourfold during hyperoxia, assisting in EC accumulation. The stimulus for H^+ secretion by renal tubule cells is presumably PCO_2, which acts by lowering IC pH.

Another mechanism of acid-base regulation which has gained some acceptance for air-breathing species is the buffering of protons by dissolution of the $CaCO_3$ exoskeleton. This would release HCO_3^- otherwise unavailable from external sources due to the loss of branchial contact with water. Although there has been some indirect evidence to support this from studies of emersed FW crayfish (hemolymph Ca^{2+} increases in proportion to base buffer, Taylor et al. 1987), there has been no direct evidence in terrestrial FW species. Any role of the kidney in acid-base balance also remains untested, although, considering the pressure on terrestrial crabs to conserve water (see Sect. 8.4), it is unlikely that this could be a significant route for acid-base excretion. Another unknown is the extent to which aerial species are capable of branchial exchange, since many of them do retain water in the branchial chamber and could potentially employ the ionic mechanisms diagnostic of water-breathers.

5.2 Intracellular Acid-Base Balance

5.2.1 Intracellular Acid-Base Profile

In the past 10 years, several authors have measured the IC pH (pH_i) in tissues of FW crustaceans (primarily crayfish) using electrodes or the distribution of the weak acid DMO (5,5-dimethyloxazolidine-2,4-dione). These studies conclude that the pH_i of soft body tissues (muscle, nerve) is around 0.2–0.6 pH units acid relative to the hemolymph (Moody 1980; Gaillard and Malan 1983, 1985; Gaillard and Rodeau 1987; Wheatly et al. 1991). Within an animal, individual tissue types can vary in the pH_i gradient they maintain with the EC pH (pH_e). Gaillard and Malan (1983) and Wheatly et al. (1991) both reported a smaller gradient between the pH_e and nerve or heart (0.2 pH units) than with muscle (0.5 pH units) in the FW crayfish. The absolute pH_i can also vary between tissue types (Wheatly et al. 1991). For example, fast muscle has a lower pH_i than slow muscle. As in vertebrates, there is a fairly good inverse correlation between pH_i and IC buffering potential (β), and a positive correlation between β and anaerobic potential. For example, fast muscles that fatigue rapidly have a higher β. In one study which has attempted to measure the pH_i of crustacean cuticle using the DMO technique, a value 0.3 pH units *above* the pH_e was reported (Wheatly et al. 1991). However, there are problems in estimating fluid compartmentalization in the carapace, on which the DMO method relies, and so these measurements should be regarded as preliminary. The carapace contains a large fluid reservoir, only part of which is penetrated by inulin. However, ultrastructural studies suggest that the IC volume is minimal, which means that there must be another EC compartment which is not inulin-penetrable. As for the remaining acid-base parameters, the IC PCO_2 has not been measured in FW crustaceans and is assumed to be equal to venous PCO_2 in the application of the DMO technique. The IC HCO_3^- concentration is typically calculated via the Henderson-Hasselbalch equation and thus reflects the pH_i and PCO_2 values. So, for example, reported values for control crayfish tissues (Wheatly et al. 1991) vary from 2 m*M* in muscle to 16 m*M* in carapace. IC acid-base parameters in FW crustaceans are in general agreement with those reported in marine species.

5.2.2 Typical Intracellular Acid-Base Disturbances and Compensatory Responses

Much of the early work on the whole-animal acid-base balance, in which the internal sampling was restricted to the EC fluid (ECF), hypothetically attributed the acidic/basic equivalent deficiencies/excesses to the IC compartment. For example, on acid exposure of crayfish (Wood and Rogano 1986), the massive influx of acidic equivalents was assumed to be buffered in the IC fluid (ICF) or exoskeleton. Recent advances in technology, in particular, the use of the DMO weak acid distribution technique, have made it possible to examine the ICF compartment under many of these environmental challenges.

Temperature

The effect of temperature on the IC acid-base status has been investigated in the FW crayfish by Gaillard and Malan (1985), who reported alphastat regulation in both the hemolymph and various soft body tissues. Depending on the temperature range, regulation was achieved by two alternate mechanisms, either ventilatory control of PCO_2 or ionic transfers across the cell membrane. Between 6 and 13 °C, regulation was due to ventilatory control of PCO_2, as in air-breathing poikilotherms, and pH_i adjustment occurred in all body fluid compartments; above 13 °C, no further increase in hemolymph PCO_2 was possible, probably due to the limitations of water-breathing, and animals built up an EC respiratory alkalosis. This was compensated only at the IC level in the heart and nerves via ionic regulation; the muscle pH_i was not restored within the 48 h time period of the experiment, since the pH_i regulation mechanism in muscle has slower kinetics than in other tissues.

Hyperoxia/Hypercapnia

Two crayfish studies have focused on pH_i regulation in response to increased EC PCO_2 elicited either by hyperoxia or hypercapnia. Gaillard and Malan (1983) studied the effect of a 48-h exposure to extreme hyperoxia (700 torr inspired PO_2) or 24 h hypercapnic exposure (6 torr PCO_2). In both cases, the EC respiratory acidosis remained uncompensated. Even so, the pH_i recovered in the heart, nerve, and muscle tissue, whereas the abdominal muscle pH_i remained depressed. The latter may reflect the lack of EC compensation since the acidic equivalent extrusion mechanism has been shown to depend on the EC pH and HCO_3^- (Rodeau 1982). Alternatively, it may again reflect the slower kinetics of pH_i regulation in muscle cells. Compensation probably occurred through the extrusion of acidic equivalents out of the cells since the strong ion difference increased. The similarity between the steady-state responses of pH_i under these two conditions (i.e., hyperoxia, hypercapnia) indicates that the pH_i regulation is essentially controlled at the cellular level.

Wheatly et al. (1991) reexamined the pH_i under 72-h hyperoxic exposure in the crayfish, this time using an inspired PO_2 of 500 torr. Crayfish were able to compensate the EC acidosis completely at this moderate level of hyperoxia (Wheatly 1989). The acid-base balance remained relatively unchanged in the heart and nerve. The skeletal muscle exhibited an initial acidosis that was primarily metabolic and was compensated within 48 h by ionic exchange and recycling acidic metabolites. At the antennal gland, lowering of the renal cell pH facilitated proton excretion into the urine; again, the acidosis was metabolically compensated. While the carapace pH_i did not change significantly, the HCO_3^- concentration increased initially and then declined, suggesting that the exoskeletal carbonate may have been translocated to assist in buffering the pH_i of soft tissues.

Aerial Exposure

Tyler-Jones and Taylor (1988) found that while the hemolymph acidosis occasioned by aerial exposure in the crayfish was compensated within 24 h, an IC acidosis persisted in muscle. Lactate accumulated in both compartments initially, but control levels were reestablished within 24 h. Since washout and reoxidation occurred on recovery in water, it is suggested that lactate was sequestered in some unexamined tissue. Discrepancies between the observed and expected hemolymph base excesses indicated that the base was mobilized from an internal source, possibly the carapace. Analysis of the IC compartment suggested that H^+ was retained inside the muscle rather than being released from the tissues with lactate. These authors suggest that the lack of compensation in the abdominal muscle may enable tissues requiring greater pH stability (heart, nerve) to regulate preferentially. This study strongly implies that the IC acid-base regulation is impaired when gill contact with water is lost. This presumably precludes the operation of the branchial ion exchanges that would appear to be the primary mechanism for the whole-animal acid-base balance. Alternative mechanisms which could be employed under these conditions are respiratory control, dissolution of internal buffer base, and IC/EC ion exchange.

Exercise

Changes in the IC acid-base balance during engagement in and recovery from exercise have not been examined in any FW species.

5.2.3 Mechanisms of Intracellular Acid-Base Balance

The mechanisms of pH_i regulation in animal cells were summarized in a recent review article by Walsh and Milligan (1989). These are (1) IC buffering, which is primarily via nonbicarbonate buffers such as phosphates and proteins and was discussed for different tissues above, (2) consumption versus production of acidic/basic metabolites, which has not been extensively researched for crustaceans, and (3) transfer of acidic/basic equivalents across the cell membrane, which is the best documented and probably the most important mechanism. So far, the most evidence exists for passive inorganic ion exchangers. An amiloride-sensitive Na^+/H^+ electroneutral exchanger has been found in the crayfish neuron (Moody 1981) and muscle (Galler and Moser 1986). An electrogenic $2Na^+/H^+$ mechanism has been identified in marine crabs in the hepatopancreas (Ahearn and Clay 1989), where it is responsible for luminal acidification, and in the gill (Shetlar and Towle 1989), where a relatively small pH difference is capable of driving a substantial apical Na^+ uptake from the dilute external salinity. A SITS- and DIDS-sensitive $Na^+/HCO_3^-/Cl^-/H^+$ electroneutral exchanger has been identified in crayfish neurons and muscle (Moody 1981; Galler and Moser 1986). Evidence is not currently available for a Cl^-/HCO_3^- exchanger, although this is evident in other animals. Also, the possibility exists for active proton ATPases and organic ion transporters.

Intracellular ion levels were monitored during the exposure of crayfish to hyperoxia and recovery (M.G. Wheatly and E.C. Vevera, unpubl. observ.). There was an initial cell swelling which was subsequently corrected. During hyperoxia, there was no evidence of a differential regulation of IC Na^+ versus Cl^-; however, on recovery, $[Na]_i$ was reduced while $[Cl]_i$ was increased, suggesting that the transmembrane electroneutral ion exchanges operated at this time to correct the pH_i and that these were more stimulated by the reinstatement of normoxia than by hyperoxia.

5.3 Relative Roles of Extracellular and Intracellular Body Fluid Compartments in Whole-Animal Acid-Base Homeostasis

In some cases, it is now possible to integrate recently obtained information on the IC acid-base balance with existing knowledge of EC changes in an attempt to delineate the varying roles of the two compartments in the whole-animal acid-base homeostasis. Based on data obtained for hyperoxic exposure of the FW crayfish (Wheatly 1989), one can calculate a total change in the EC H^+ pool of $-923\,\mu mol\,kg^{-1}$ and $-2469\,\mu mol\,kg^{-1}$ in the ICF. The 1:3 ratio between the EC:IC H^+ production suggests that the ICF is quantitatively more important in generating the whole-animal acid-base disturbance. The measured net flux of H^+ into the experimental water was approximately equivalent to the whole-animal change in the H^+ pool, confirming that this is the primary avenue for acid-base compensation. The removal of a crayfish into air (Tyler-Jones and Taylor 1988) resulted in opposing trends in the H^+ pool of ECF ($-1232\,\mu mol\,kg^{-1}$) and ICF ($+4024\,\mu mol\,kg^{-1}$), presumably because branchial ion exchange was no longer possible.

6 Molting Physiology

As arthropods, crustaceans possess an exoskeleton, mineralized primarily with $CaCO_3$, which must be molted in order to increase in size. This presents the animal with a range of physiological challenges, particularly with respect to ionoregulation (see Mangum 1985). The structural aspects of crustacean biomineralization have been extensively reviewed (Travis 1963; Roer and Dillaman 1984; Lowenstam and Weiner 1989; Simkiss and Wilbur 1989); however, the physiological mechanisms enabling biomineralization have been less well studied. Greenaway (1985) reviewed the Ca balance in crustaceans, and Wheatly (1993) recently reviewed the electrolyte regulation throughout the molting cycle specifically in the FW crayfish.

The ionic demands of molting are exacerbated in FW because the external Ca is generally limited to less than 1 m*M* compared with values of

5 m*M* in SW. In fact, one study has shown that crayfish do not occur in waters with Ca levels below 0.125 m*M* (Jay and Holdich 1981). The Ca balance during the molting cycle has been described fairly comprehensively in the FW crayfish (Greenaway 1972, 1974a,b,c; Wheatly and Ignaszewski 1990) and was outlined for intermolt animals in Sect. 4.2.

The frequency of molting will depend upon the developmental stage, sex, and environmental factors (Lowery 1988). Crayfish hatchlings molt every 6 days, and the frequency declines to once yearly in 4–5-year-old adults. Males of some species achieve two molts per year. Environmental factors which can affect molting are temperature, ion and food availability, and population density.

6.1 Calcium Regulation

This review of molting physiology will begin with a discussion of the Ca balance. In the premolt period, the hemolymph total Ca rises, but this is mainly in the form of bound Ca; the ionized Ca^{2+} value remains constant (McWhinnie 1962; Greenaway 1974b,c). The urine levels meanwhile rise, but only to 2.7 m*M*, suggesting that the renal Ca^{2+} reabsorption persists and that urinary loss is minimal (Greenaway 1974b). This high degree of Ca regulation is somewhat unexpected because the whole-body Ca is undergoing dynamic changes at this time, with resorption from exoskeletal reserves, deposition in storage areas, and substantial loss to the environmental water.

Very early on, it was observed that $CaCO_3$ concretions (gastroliths) appeared in the cardiac stomach of crayfish around 40 days before ecdysis in adults and disappeared within 2 days following ecdysis (Huxley 1879; Travis 1960; Chaisemartin 1964). It was automatically assumed that they afforded a storage site for Ca reabsorbed from the old exoskeleton. Their quantitative role in Ca storage, however, has been reassessed. In terms of the whole-animal Ca budget at the time of ecdysis, the gastroliths only store 12% of body Ca (Chaisemartin 1964; Wheatly 1990); an additional 4% is stored in the soft body tissues including hemolymph. Thus, Ca storage is minimal (16%), even though the animal is Ca-limited in its environment. About 35% of the body Ca is lost in the shed exuviae, which suggests that as much as 49% of the body Ca is lost to the environmental water in the days preceding the molt, which seems maladaptive. A whole-animal net Ca^{2+} efflux of $-800\,\mu mol\,kg^{-1}\,h^{-1}$ has been reported in premolt crayfish (Greenaway 1974b; Wheatly and Ignaszewski 1990). The route for Ca loss is presumed to be branchial. Since the rate of efflux is 25-fold greater than in intermolt and permeability to other ions only doubles at this time, there is reason to suspect that the loss is not simply by passive diffusion but may involve a $Ca^{2+} + HCO_3^-$ cotransporter or a $Ca^{2+}/2H^+$ exchanger; the mechanism is probably not active since the electrochemical gradient favors Ca^{2+} loss (Greenaway 1985).

The premolt Ca^{2+} efflux is accompanied by an excretion of basic equivalents (alternatively acidic equivalent uptake, Wheatly and Ignaszewski 1990). Respirometric measurements from that study revealed that the CO_2 production exceeded O_2 uptake by a comparable amount at this time, confirming that the acid-base transfer was a base excretion. This is what one might predict based on the equation for the resorption of cuticular $CaCO_3$ ($CaCO_3 + H^+ = Ca^{2+} + HCO_3^-$).

The molting cycle terminates in crustaceans with ecdysis or the physical shedding of the old exoskeleton. The increase in volume required for this is believed to be precipitated by water absorption, although this has only been documented in marine species (Mykles 1980). Since the crayfish is ordinarily hyperosmotic, a stimulus exists for water uptake, although the precise details of how this occurs are unknown. For example, why would the kidney not increase urine flow to restore the EC volume? Knowing that crayfish experience a 13% increase in mass at this time, one can predict the effect this would have on hemolymph dilution (36% reduction in circulating electrolyte concentration) assuming that water was taken up exclusively into the ECF. Measured reductions in the hemolymph Na^+ (Wheatly 1990) confirm this to be the case, although there appears to be some regulation of hemolymph Ca^{2+} and Cl^- levels at this time.

During postmolt, the primary ionic event is mineralization of the new exoskeleton so that the animal can once again become protected and resume its regular activities. The Ca^{2+} originates from the external water or from stores within the body such as gastroliths. Calcium budgets constructed for the FW crayfish (Wheatly 1990) suggest that the gastroliths are relatively unimportant in that they contribute only 12% of the intermolt whole-body Ca content. Furthermore, they persist for the first 2 days postmolt (Willig and Keller 1973), suggesting that they are not involved in the initial hardening. In our laboratory, we have also observed that in some aberrant animals, which were incapable of hardening, the gastroliths remained intact after several days, indicating that the absorption of external Ca^{2+} may be a prerequisite for gastrolith dissolution. Chaisemartin (1964), however, reports that gastroliths are remobilized after 8 h and are important in initial hardening. This discrepancy needs to be resolved. In the FW land crab *Holthuisana*, the hemolymph serves as a Ca^{2+} storage depot (Sparkes and Greenaway 1984). The hemolymph becomes cloudy white and contains microspheres of $CaCO_3$.

Another debatable issue is whether crayfish eat their cast exoskeletons. The Ca budget (Wheatly 1990) indicates that the exoskeleton contains about 35% of the intermolt whole-body Ca. Animals that are not allowed to reingest the exuviae reabsorb only 50% of their Ca requirement during the 2 weeks following ecdysis and so would require additional uptake to attain intermolt values. Thus, ingesting the exoskeleton would seem highly adaptive in conserving Ca. In laboratory crayfish, we have observed this on occasion, but only rarely; however, these animals may be stressed in captivity. There is no consensus as to whether this occurs routinely in the wild.

The most important source of Ca is uptake from external water (around 50% of the total Ca requirement, or more if the animal refrains from eating its exuviae) which commences at a rate of $+2000\,\mu mol\,kg^{-1}\,h^{-1}$ and is accompanied by a basic equivalent uptake (Wheatly and Ignaszewski 1990). Flux rates decline exponentially with time but remain significant even after several days. Calcification may continue at a reduced level for some time. The hemolymph Ca value meanwhile remains remarkably constant (Wheatly 1990) at essentially intermolt levels, which is surprising considering the Ca dynamics occurring at this time and suggests that there is strict Ca regulation.

The mechanism for Ca^{2+} uptake has been somewhat elusive and may involve a combination of different mechanisms. The rate-limited curves obtained for Ca^{2+} uptake in the crayfish by Greenaway (1974c) reported a K_m of $0.13\,mM$, a saturation of $0.4\,mM$, and a K_{max} of $2\,\mu mol\,g^{-1}\,h^{-1}$ at 10 °C. All features are typical of ion-transport ATPases, and it was concluded that uptake may involve a branchial Ca^{2+}-ATPase along the lines of the model proposed for FW fish by Flik et al. (1985). However, recently in my laboratory (Wheatly 1990), we were unable to demonstrate any increase in Ca^{2+}-ATPase activity in gill microsomes of postmolt crayfish. The only increase occurred in the cuticular hypodermis. Thus, it would appear that the Ca^{2+} influx mechanism may be passive, as has been elucidated for marine decapods (Cameron 1989b); it is probably electroneutral and may involve a $Ca^{2+}/(2H^+$ or $2Na^+)$ exchanger (Greenaway 1985). This area awaits additional research.

Indirect information about Ca^{2+} uptake mechanisms have been gleaned from experiments on crayfish in my laboratory (Wheatly and Gannon 1993), which indicate that calcification is impeded (50–60%) if Na^+ or HCO_3^- are removed from the external solution. If the postmolt crayfish are placed in Ca^{2+}-free medium, they remain in Ca^{2+} balance, suggesting that a passive Ca^{2+} efflux is not possible at this time. If, after 4 days, Ca^{2+} is reintroduced into the external medium, the crayfish begin to calcify but only at rates typical of a 5-day postmolt crayfish, which are insufficient to accomplish adequate hardening. This strongly suggests that the opportunity for maximal calcification is preprogrammed into the initial postmolt period and that it cannot be temporally shifted. Both the mechanisms and regulation of postmolt calcification remain subjects of immense interest.

For calcification to be accomplished, HCO_3^- must also be available. While this could theoretically originate from respiratory CO_2, we have fairly convincing evidence (Wheatly and Gannon 1993) that the external water is once again the major source, since the basic equivalent uptake is stoichiometrically equivalent to a deficit in CO_2 production at this time, and calcification stops in HCO_3^--free medium. In the process of converting HCO_3^- to CO_3^{2-}, H^+ is produced as a by-product, which can be eliminated at the gill in exchange for Na^+. The mechanism for HCO_3^- uptake is unknown, but it may be indirectly linked with the Ca^{2+} uptake mechanism since the time courses are similar.

The deposition of $CaCO_3$ mineral in the new cuticle has been extensively studied ultrastructurally (reviewed by Simkiss and Wilbur 1989). Mineral first appears around the pore canals and then throughout the chitin-protein fibrillar network. Recent evidence suggests that there is a physiologically distinct subdivision of the EC fluid in the carapace (termed carapace fluid) that maintains a relatively alkalotic pH (0.3 pH units above hemolymph pH, see Sect. 5.2.1) purportedly to prevent the mineral from redissolving (Wheatly et al. 1991). The existence of this compartment remains somewhat controversial and merits further research.

6.2 Regulation of Other Electrolytes

Crustaceans molting in FW are faced with unique ionoregulatory problems associated with the postmolt regulation of Na^+ and Cl^-. As outlined above, the major hemolymph ions experience dilution due to the influx of FW to assist in shedding. Thus, in addition to taking up Ca^{2+} and HCO_3^- for calcification, they are faced with a net Na^+ and Cl^- uptake to offset the hemodilution. Uptake commences at rates of +400 and +800 $\mu mol\, kg^{-1}\, h^{-1}$, and the ion balance is essentially reestablished within 2 days (Wheatly 1993). The mechanisms involved are no doubt the same ones involved in hyperionic regulation in intermolt animals and are probably located on the gills. Control levels can be reestablished even in Ca^{2+}-free medium; however, when Ca^{2+} is reintroduced, there is a secondary uptake of Na^+, suggesting again that the Na^+ and Ca^{2+} uptake mechanisms are somehow linked. The role of the antennal gland in postmolt ion regulation has not been investigated.

6.3 Effect of Ambient pH

Because HCO_3^- is integral to the calcification process, the ambient pH will have profound effects on calcification by virtue of its effect on CO_2 equilibria. Furthermore, as outlined above and below, FW environments have a reduced buffering potential compared with SW and therefore will experience more dramatic changes in pH. While some of these pH changes may be naturally occurring, most are anthropogenically generated, such as acid rain and the discharge of alkaline coal ash wastes. France (1987a) reported a decreased carapace rigidity and Ca content in crayfish under acid stress in the wild. Malley (1980) found that while intermolt crayfish could survive experimental exposure to acid pH (4), there was high mortality among postmolt animals. The latter group progressed more slowly through the molt cycle and calcified at a reduced rate. The author offered the following as possible explanations: (i) acid pH results in lowered ambient HCO_3^-, which has been shown above to reduce calcification, (ii) high external H^+ may impede excretion of H^+ produced as a by-product of calcification, or (iii) H^+ may have a direct effect on the ion transport systems involved in calcification. Although a low pH impairs calcification, it is unlikely that this is the direct cause of death, which is thought to result from ion or acid-base imbalance (see Sect. 9.2). Shaw (1960b) has shown that Na^+ uptake is inhibited below pH 6.0.

Recently, Zanotto and Wheatly (1993a,b) reexamined whole-animal postmolt electrolyte fluxes in acid (pH 5.1) or alkali (pH 9.1) water in the FW crayfish. In acid water, there was a 65% reduction in Ca^{2+} and basic equivalent uptake and a significant reduction in Na^+ uptake attributable to the mechanisms outlined above. In alkali water, there was a 20% reduction in Ca^{2+} uptake, although the base uptake increased initially, suggesting that the two are not necessarily tightly coupled; Na^+ experienced a net efflux. Thus, postmolt calcification and restoration of the Na^+ balance were impaired in both acid and alkaline waters; the Cl^- uptake, however, was unaffected in either treatment. Many of these observed changes may be alternatively explained by changes in the ion/acid-base balance in the hemolymph, and we are currently investigating this aspect.

This series of measurements has recently been repeated in decarbonated water (Zanotto and Wheatly 1990) in order to avoid the complicating effects of varying PCO_2 between pH treatments. In decarbonated water, postmolt fluxes of Ca^{2+}, basic equivalents, Na^+, and Cl^- were the same in all treatments (i.e., no observed effect of pH). However, calcification in neutral decarbonated water was 70% reduced compared with neutral non-decarbonated water. This indicates that calcification is more dependent on the ambient HCO_3^- than on pH per se. Thus, the results in acid non-decarbonated water may be attributed to the effect that acidification has on lowering the ambient HCO_3^-. Reduced calcification in alkaline non-decarbonated water requires some other explanation, such as the direct effect of an increased pH on the ion transport mechanisms, since the ambient HCO_3^- is not limited under these conditions.

The acidification of environmental waters is often associated with elevated concentrations of trace metals; these result either from atmospheric deposition or leaching from soils. The latter is especially true of aluminum, which can be toxic to the aquatic fauna. While the combined effects of acidity and high Al have not been systematically tested on intermolt decapods, Malley and Chang (1985) observed their combined effect on postmolt Ca^{2+} uptake in the crayfish, *Orconectes virilis*. They concluded that Al toxicity constitutes a small additional stress to Ca^{2+} uptake in softwater acid lakes (pH 5–6). In nonacidified water, Al ($200\,\mu g\,l^{-1}$) had no effect on the Ca^{2+} uptake. In acid water, the calcification was reduced to approximately 30% of that in neutral water, and Al at concentrations of

200, 500, or 1000 $\mu g\,l^{-1}$ in acid waters reduced calcification to 20% of the level in the nonacidified medium. The degree of inhibition was not related to the concentration, and this may be explained by the limited concentration of monomeric species which are present in spite of increases in the Al concentration. A lack of Al toxicity at pH 7 may be due to changes to less toxic chemical species or the apparent lower solubility of $AlCl_3$. The collapse of crayfish populations at pH 5.5 may be exacerbated by Al effects on postmolt uptake; however, the causes of failure are manifold.

7 Gas Exchange: Ventilation and Circulation

The most recent general review articles on gas exchange in crustaceans are McMahon and Wilkens (1983), Burggren and McMahon (1988b), McMahon and Burggren (1988), and Cameron (1989c). In this section, attention will be focused on factors which challenge gas exchange in fresh water; adaptations which enable many FW species to tolerate aerial exposure are covered in the following section (Sect. 8).

7.1 Branchial Morphology

In crustaceans, the primary organ of aquatic gas exchange is the gill. Three types of gills are recognized: (i) pleurobranchs, which originate from the pleural wall, (ii) arthrobranchs, which originate from the articular membrane between the body wall and appendage; and (iii) podobranchs, which originate from the coxa of the appendage (McLaughlin 1983). The precise branchial formula varies among species but is typically 13–18 pairs in crayfish and 9 pairs in crabs. The gill itself consists of a central axis with afferent and efferent vessels and a series of lateral branches, the arrangement of which varies between the infraorders Brachyura and Astacidea. The gill surface area is approximately 5 $cm^2\,g^{-1}$ in FW crayfish (Bergmiller and Bielawski 1970), which is comparable to values measured in marine crabs and thus does not ostensibly appear to reflect the ionic status of the medium. Crabs typically possess phyllobranchiate gills in which the lateral branches are flattened lamellae, usually set in two rows. Crayfish possess trichobranchiate gills, in which the branches are filamentous and arranged in several series around the central axis. The morphology of the crayfish gill originally described by Huxley (1879) and Bock (1925) has been reexamined by Fisher (1972), and the functional morphology was reinterpreted by Burggren et al. (1974). Inside each filament in the crayfish, there is a transverse septum running from the base almost to the tip, dividing the afferent and efferent hemolymph. There are small openings in this septum to allow some crossover, possibly allowing a change in the effective length of the filament. This could be employed to balance the animal's respiratory and more importantly ionic requirements. Water flows parallel to the afferent stream (i.e., cocurrent) and counter to the efferent stream; however, the latter has a larger cross-section, so that countercurrent exchange is maximized. In the Brachyura, the lamellate gill structure is slightly more elaborate since the flat platelets extend to either side of a supporting arch, and hemolymph channeling is more effective since the afferent and efferent vessels lead to either end of the lamellar space and provide exclusively countercurrent flow (Hughes et al. 1969). The diffusion distance is typically 5 μm in FW brachyurans and astacideans, agreeing again with values for marine species. The gills are housed in the gill chamber formed by lateral extensions of the carapace (the branchiostegites), and they are ventilated by the beating motion of the scaphognathite.

7.2 Ventilation

Ventilation of the branchial cavity is effected by oscillations of the scaphognathite, which is the flattened exopodite of the second maxilla. Intrabranchial pressure recordings made in crayfish (McMahon et al. 1974; Taylor and Wheatly 1980) demonstrate a biphasic wave-form generating subambient pressure (around 1 cm H_2O). This aspirates water into the hypobranchial space (via holes at the bases of the legs), across the gills, into the epibranchial space, and out of the exhalant openings. In crayfish, due to the arrangement of the afferent and efferent hemolymph vessels outlined above, countercurrent flow between water and hemolymph can only occur in part of the system, and as such, gas exchange is less effective than in brachyurans. Forward pumping predominates at gill ventilation rates ($\dot{V}w$) of approximately 96 ml kg^{-1} min^{-1} (Wheatly and Taylor 1981). Ventilation is predominantly bilateral in crayfish; the unilateral pumping and respiratory pauses so common in brachyurans do not appear to be regular ventilatory events. However, periodically, the scaphognathite reverses the pressure gradient. While brachyurans may reverse it for extended periods, during which time the water flow changes direction, crayfish reversals are restricted to single beats and only occur rarely. The function of these reversals has been long debated (see McMahon and Wilkens 1983). Possible explanations are to clear particulate matter from the gills or to stir up poorly ventilated areas in the gill chamber.

7.3 Circulatory Morphology

The "open" circulatory system of decapod crustaceans has been reviewed in the past by Maynard (1960), Taylor (1982), McMahon and Wilkens (1983), and Burggren and McMahon (1988b). The central pump is the dorsal, single-chambered heart, which is situated within the pericardial sinus. The latter receives oxygenated hemolymph from the gills via the branchiopericardial veins and also from the inner wall of the branchiostegites. The hemolymph then enters the heart via three pairs of ostia whose opening is regulated by valves. The heart beat originates in a neurogenic pacemaker, which is the cardiac ganglion located inside the dorsal wall. The heart rate is modified by cardioregulatory nerves originating in the subesophageal ganglion and by hormones (peptides and biogenic amines) released from the pericardial organs. The hemolymph is then ejected into several arteries for distribution to the tissues. Vessels, structurally and functionally analogous to the capillary bed of vertebrates, have been described; however, it is uncertain whether the hemolymph distribution can be altered by changes in peripheral resistance. Following this, the hemolymph drains into spaces called lacunae that collect into a network of sinuses. These perfuse the gills and branchiostegal membranes before returning via the branchiopericardial vessels to the pericardium. Thus, the gills are essentially perfused by venous return to the heart. Each branchial filament contains an afferent and efferent vessel that are separated by a continuous septum in the majority of crabs. As outlined earlier, this septum is perforated in the crayfish (Burggren et al. 1974) and is absent altogether in the FW land crab *Holthuisana* (Taylor and Greenaway 1979). Although valves have been observed in the lamellae of a marine crab (H.H. Taylor and Taylor 1986), it is generally uncertain whether crustaceans actively control hemolymph flow within individual lamellae.

7.4 Cardiovascular Function

Certain cardiovascular variables have been determined for FW crustaceans. The heart rate (fH) in crayfish (at 15 °C) is approximately 80 beats min^{-1}, with a rate of gill perfusion ($\dot{V}b$) determined by the Fick principle of 83 $ml\,kg^{-1}\,min^{-1}$, and a cardiac stroke volume of around 1 ml (Taylor and Wheatly 1980). The cardiac output is high compared with other groups of animals, and this is no doubt correlated with the lower O_2 carrying capacity of crustacean hemolymph, as well as the relatively large EC volume in open hemolymph systems. All of these values are consistent with values in marine species. The heart operates as a force suction pump, aspirating hemolymph during diastole. In marine species, the systolic and diastolic intracardiac pressures are approximately 10 and 3 mmHg, respectively, and the pericardial sinus pressure is virtually 0. Since the pressure differential across the branchial vasculature is only around 2 mmHg, most of the fall in pressure through the circulatory system occurs in the systemic tissue (Blatchford 1971). Branchial perfusion may be assisted by subambient pressures which are generated in the branchial chamber during ventilation and transmitted across the thin lamellar surface. Single reversed beats may serve to force hemolymph from the gill channels, providing a sudden pulse to the pericardial cavity.

7.5 Rate of O_2 Uptake

Quiescent rates of O_2 uptake ($\dot{M}O_2$) for FW crustaceans are fairly similar to those reported for marine species at comparable temperatures and activity levels (McMahon and Wilkens 1983). Typical values for crayfish at 15 °C are 12 $\mu mol\,kg^{-1}\,min^{-1}$ (Taylor and Wheatly 1980). $\dot{M}O_2$ is affected by a number of environmental and endogenous factors. The temperature dependency of the crustacean $\dot{M}O_2$ is reflected in values for Q_{10} of 2–3 (Taylor 1981). The environmental O_2 tension will also affect $\dot{M}O_2$, and the response of individual species will generally reflect the O_2 content of their natural habitat. For example, the crayfish *Pacifastacus leniusculus*, which is found in waters of low temperature and high O_2 content, does not regulate $\dot{M}O_2$ over a wide range of O_2 tensions (Moshiri et al. 1970). However, other species endemic to more stagnant bodies of water, such as *Procambarus simulans*, can regulate $\dot{M}O_2$ down to low external O_2 levels (Larimer and Gold 1961). One study has reported that cave-dwelling crayfish have a lower $\dot{M}O_2$ than surface dwellers, perhaps as an adaptation to the reduced food supply (Dickson and Franz 1980). However, a recent reevalulation of this issue (Gannon et al. 1990) failed to show any difference between the $\dot{M}O_2$ of epigean and hypogean crayfish. There are a multitude of other abiotic factors (e.g., salinity, latitude, pollutants) which can affect $\dot{M}O_2$, and these were recently reviewed by Vernberg (1983).

As for the effect of biotic factors on $\dot{M}O_2$, again these are much the same as for other crustaceans. Activity (see Sect. 7.6.2) results in an aerobic metabolic scope which is approximately four times the routine level in the crayfish (Rutledge and Pritchard 1981). The metabolic processes surrounding ecdysis similarly double $\dot{M}O_2$ (Wheatly and Ignaszewski 1990), and there are many other physiological activities such as egg production, feeding, parasitism, and diurnal rhythms which can all affect $\dot{M}O_2$.

7.6 Challenges to Gas Transport

7.6.1 Hypoxia

Crustaceans experience periods when the routine O_2 delivery cannot meet an increasing demand. For FW crustaceans, this is most noticeable when they routinely encounter environmental hypoxia. As outlined above (Sect. 7.5), many are able to regulate $\dot{M}O_2$ down to low external levels using a sophisticated suite of physiological mechanisms. Both acute and chronic studies have been undertaken.

Wheatly and Taylor (1981) studied the effect of progressive hypoxia on gas exchange in the European crayfish, *Austropotamobius pallipes*. Inspired PO_2 was reduced to 30 torr over a period of approximately 12 h. They discovered that crayfish could maintain $\dot{M}O_2$ down to a critical PO_2 of around 40 torr. The heart rate progressively slowed, but the hemolymph flow was maintained by an increase in stroke volume. By 50 torr, the gill ventilation had increased almost threefold, due to an increase in respiratory frequency combined with increases in the mean hydrostatic pressure recorded in the branchial chamber and the mean amplitude of the pressure pulse. The $\dot{M}O_2$ value rose as the branchial ventilation increased, indicating that there is a cost to the hyperventilation. While the effectiveness of the removal of O_2 from water was maintained, there was evidence to suggest that the ability of the respiratory surface to transfer O_2 ($\dot{T}O_2$) was improved. The hyperventilation generated a hemolymph alkalosis, which had the effect of increasing the HCy O_2 affinity (P_{50} decreased from 8 to 4.5 torr). This enabled the arteriovenous O_2 content difference to be maintained even though pre- and postbranchial tensions had both decreased by 50%. Below 50 torr, crayfish were unable to sustain the hyperventilation, and the effectiveness of O_2 removal into the blood decreased, so that $\dot{M}O_2$ fell. In other words, this is the point at which the gas exchange system fails to continue to provide sufficient O_2 to maintain the normoxic level of aerobic metabolism. There was a switch to anaerobic metabolism with an incipient acidosis.

McMahon et al. (1974) examined the effect of long-term (10-day) exposure to inspired PO_2 of 50 torr in the crayfish, *Orconectes virilis*. The initial response was a hyperventilation (sixfold increase in $\dot{V}w$) and a 50% increase in heart rate, allowing $\dot{M}O_2$ to be maintained from O_2-depleted water. However, over a period of 8 days, the respiratory and circulatory pumping rates had all returned towards prehypoxic levels, although $\dot{V}w$ remained somewhat (125%) elevated. Presumably, in the long term, adaptive changes in other parameters increase the efficiency of gas exchange. For example, realignment of the branchial appendages could reduce the water flow resistance. Acclimative changes could also occur in the circulatory system, such as the opening of previously poorly perfused gill filaments or changes in the O_2 combining characteristics of the respiratory pigment. Below 30 torr, the circulatory and respiratory pumping rates rapidly declined, and death resulted.

The nature of these adaptive changes was investigated further in two papers by Wilkes and McMahon (1982a,b). In the first, the crayfish *Orconectes rusticus* was exposed to 6 days of hypoxia (50 torr). The pattern of responses was very similar to that observed by McMahon et al. (1974). The initial transitory hyperventilation and perfusion may have constituted an avoidance response. The extent of oxygen extraction during hypoxia was well below normoxic values, and so maintenance of the $\dot{M}O_2$ could not have resulted solely from an increased O_2 supply. Once again, during the latter 72 h of hypoxia, both scaphognathite pumping and cardiac output were significantly below maximum levels, although they remained above pre-hypoxic levels. This served to increase O_2 delivery to the branchial chambers and tissues and at the same time assisted in the removal of CO_2 from the hemolymph, effecting a respiratory alkalosis. This was associated with a base deficit but was never fully compensated. The elevated hemolymph pH increased the HCy O_2 affinity via the Bohr effect, allowing the postbranchial hemolymph to achieve a greater degree of saturation at lower O_2 tensions. The increased affinity also maintained the pressure gradient across the gills by removing O_2 from the solution in the hemolymph. Utilization of the venous O_2 reserve was not a possibility in this case because little O_2 is held in reserve under normoxic conditions. Increased gill perfusion rapidly removes the oxygenated hemolymph from the gills as well as increasing the effective surface area, potentially by reducing the magnitude of the afferent/efferent shunt (described in Sect. 7.1) in filamentous gills.

In the second study (Wilkes and McMahon 1982b), the hemolymph levels of electrolytes and protein, acid-base status, buffer value, Bohr value, and O_2 affinity were monitored in crayfish before and after 3.5 weeks of hypoxic exposure. The HCy concentration did not increase. In the case of FW organisms, this is perhaps desirable because increased levels could raise the colloid osmotic pressure and exacerbate existing problems with the water balance. Also, it could elevate the hemolymph viscosity and thus the energy required for perfusion. However, the O_2 affinity was elevated partly (50%) by a Bohr shift due to the respiratory alkalosis, in addition to a proton-independent increase in affinity due to an unidentified cation.

Having discussed the physiological responses of the crayfish to a reduced environmental PO_2, it should be noted that, if allowed access to air, these animals will voluntarily migrate out of hypoxic water and commence air breathing (Taylor and Wheatly 1980; McMahon and Wilkes 1983). Interestingly, these behavioral adaptations are elicited at an inspired PO_2 of between 30–40 torr which is coincidentally the critical PO_2 for $\dot{M}O_2$ in water. Thus, it would appear that animals will choose to breathe air rather than remaining in O_2-depleted water and switching to anaerobic metabolism. The physiological consequences of the behavioral switch to air-breathing are discussed below.

Changes in PO_2 appear to be detected by O_2 receptors which initiate the ventilatory reflex response. Massabuau et al. (1980) reported that an abrupt augmentation of PO_2 from 40 to 400 torr at the entry to the branchial cavity in the crayfish resulted in a ventilatory decrease within 15 s. They concluded from the ventilatory response latency that the chemoreceptors initiating the ventilatory reflex are situated peripherally in the gills and/or in the branchio-cardiac trunks. Massabuau and Burtin (1984) further investigated the role of these peripheral receptors in ventilatory and circulatory responses by exposing crayfish to brief hyperoxic exposure. The response latencies increased with the increase in inspired PO_2 (PIO_2), whereas the time required for the hemolymph to reach the heart and central nervous system (CNS) remained constant and independent of PIO_2 and $\dot{V}w$. So they concluded that the receptors are located in the gills. The same group have also provided evidence for CO_2 chemoreception in the gills (Massabuau and Burtin 1985).

7.6.2 Exercise

Exercise is another situation in which the O_2 demand typically outstrips the supply. Physiological adaptations have been reviewed by McMahon (1981), Taylor (1982), and McMahon and Wilkens (1983). A few of the existing studies have been undertaken on FW decapods. The mode of exercise exhibited by FW crayfish is a combination of tail flipping and walking. The most noticeable response to exercise is a marked increase in $\dot{M}O_2$, termed the aerobic metabolic scope. At 10 °C this constitutes a fourfold increase over resting values in a variety of crayfish species (Hassall 1979; McMahon and Hassall 1979; Rutledge 1981), which is in line with the values for other crustaceans. However, Rutledge and Pritchard (1981) reported a maximum aerobic scope of ten times the resting rate in the crayfish *Pacifastacus leniusculus* at 20 °C, confirming observations made in fish that the aerobic metabolic scope increases with temperature to a critical value (which is 20 °C in crayfish) and then declines steadily. The scaphognathite rate increased immediately with exercise, suggesting some kind of proprioceptive control mechanism, and this caused a tripling in the water pumping rate associated with a reduction in percentage O_2 extraction. Pumping became continuous, and unilateral/bilateral pauses and reversals ceased to occur. This large hyperventilation initially created ion loss problems in the crayfish. The passive loss decreased with time, suggesting that the branchial exchange is progressively limited. The cardiac pumping also increased primarily due to increases in stroke volume. Both the post- and prebranchial PO_2 decreased during activity, and a sharp Bohr shift allowed incursion into the venous reserve, enhancing O_2 release to the tissues. The hemolymph acidosis originates from CO_2 (70%) and metabolic acid (30%) accumulation. Even the maximal aerobic scope is generally inadequate to meet the energetic requirement, and animals resort to anaerobic metabolism, leading to lactate accumulation (Rutledge 1981). In considering where the limitation on O_2 uptake occurs, the scaphognathite pumping frequencies reach a maximum early in enforced exercise and decrease at increased temperatures (Rutledge 1981). Thus, there may be limitations in the O_2 supply to the gills. There is also evidence for a perfusion limitation, since the cardiac pump appears to reach maximum levels. Overall, crayfish show physiological responses far more similar to those of running crustaceans than to those of swimmers. Recovery is generally an extended process in crustaceans, although the specifics have not been studied in FW crayfish.

8 Aerial Exposure

As outlined above, inland waters, especially ponds, can dry up at times of drought, exposing the inhabitants in air. This will present problems with gas exchange and ion and water balance and may be solved either by physiological adaptations or behavioral modifications. The respiratory adaptations for air-breathing have been investigated in FW decapods ranging from those that live primarily in water to those that more routinely encounter aerial exposure.

8.1 Facultative Air-Breathing: Respiration, Circulation, and Gas Transport

Some crayfish, for example *Orconectes rusticus*, exhibit a ventilatory pattern intermediate between water- and air-breathing which was described by McMahon and Wilkes (1983) and termed the "emergence response". In severely hypoxic water (below 30 torr) crayfish will surface and voluntarily ventilate one or both branchial chambers with air while remaining in water. To do this, the animals must expose the inhalant openings at the base of the limbs and are seen rolling at the surface alternately ventilating right and left branchial chambers. An increasing ambient PCO_2 (13 torr) increased the duration but not the frequency of the response. During emergence, the respiratory frequency (fsc) decreased while fH increased. Upon resubmergence, fH rapidly returned to preemergence values, but fsc remained reduced for several minutes. Emergence increased the oxygenation of the gills, as evidenced by a doubling in the postbranchial (arterial) O_2 tension (PaO_2). However, the respiratory and cardiac reflexes are so rapid that they are thought to be mediated by air exposure per se and are initiated by mechanoreceptors sensitive to the change in density of the medium. Chemosensory input (central O_2 receptors) may be involved in the subsequent characteristics of the emergence episode. The advantage of this periodic pattern of emergence over complete removal into air is that contact with the water is retained for the purpose of CO_2 excretion and maintenance of the ion/water balance.

In crayfish species of the northern hemisphere, only limited terrestrialization appears to have occurred. Even so, the aquatic species found there do venture onto land in search of food or to colonise new bodies of water. Taylor and Wheatly (1980) found that a migration into air can be precipitated if the water inspired PO_2 drops below 42 torr. The same study also showed that on exposure in air at 15 °C for 3 h, the crayfish *Austropotamobius pallipes* were able to maintain $\dot{M}O_2$ (after an initial reduction due to disturbance), fsc, fH, and $\dot{V}b$, even though $\dot{V}w$ was reduced to 5%. This suggests either that the scaphognathite is less effective as an air pump or that ventilation is reduced to minimize the evaporative water loss, which is one of the primary problems in terrestrial species. The hemolymph gas tensions revealed that the crayfish were internally hypoxic, with the PaO_2 reduced to 33% and the arterial O_2 content (CaO_2) to 45% of the submerged values. $\dot{M}O_2$ was maintained by drawing on a venous reserve of O_2, reducing the prebranchial PO_2 (PvO_2) to 15% of the submerged value. The transfer factor dropped from 0.13 µmol torr^{-1} in water to 0.09 in air, suggesting that either the ventilation in air is inadequate or that there is a large barrier to diffusion, both of which could be explained by a partial collapse of the gills in air. At the same time, there was a marked hemolymph acidosis of respiratory (CO_2 accumulation) and metabolic (lactate) origin. The lactate accumulates during the initial hour of exposure when water draining from the branchial chambers leaves the gills clumped and consequently ineffectual at gas exchange until they dry out. On a return to water after 3 h, all these parameters were quickly restored to settled levels. There was an initial hyperventilation, presumably to replenish the venous reserve, but no apparent repayment of an O_2 debt.

These preliminary findings were extended in a study of the long-term (24 and 48 h) effects of aerial exposure in the same species (Taylor and Wheatly 1981). Again, $\dot{M}O_2$ was maintained at preexposed levels for the first 24 h. The initial hemolymph acidosis was compensated within 24 h by an elevation of bicarbonate buffer and reduction in circulating lactate, due either to progressive reoxidation to pyruvate or sequestration in the tissues. Since the gills were no longer in contact with water, acid-base regulation through ion exchange could not occur, and so presumably the HCO_3^- originated from an internal source, possibly the $CaCO_3$ skeleton. With both Pa and PvO_2 initially low, the arterio-venous O_2 content difference (CaO_2-CvO_2) was initially close to zero; however, within 24 h both CaO_2 and CvO_2 had returned to the levels in submerged animals. These changes were explained by the effect of the Bohr shift on HCy O_2 binding. The Bohr factor was large (closely resembling levels in other air-breathing crustaceans) and may represent a preadaptation to confer some resistance to the initial pH changes associated with moving into air. On return to water after 24 h in air, $\dot{M}O_2$, fsc, fH, and $\dot{V}w$ were elevated initially, and there was a temporary alkalosis; however, all parameters were returned to control levels within 8 h. The rapidity of recovery is attributed to base excretion across the gills. Lactate reappeared in the blood as it was washed out of the tissues. While the crayfish appeared to tolerate 24 h of aerial exposure with no ill effects, a longer exposure presented problems. After 48 h in air, $\dot{M}O_2$ and fH were significantly reduced, and ventilation became intermittent. Why compensation should break down after 24 h remains a mystery; however, it could result from the ancillary problems facing air-breathers of evaporative water loss and loss of an aquatic route for ion exchange. There was a secondary acidosis and reduction in the arteriovenous content difference. This species can remain in moist air for approximately 1 week, although the animals become increasingly inactive.

In the southern hemisphere, terrestrialization may have proceeded further in the Parastacid crayfish than in either the Astacidae or Cambaridae of the northern hemisphere. Here, an extensive use of burrows enables many species to attain considerable independence from water. Swain et al. (1988) compared the branchial morphology of two Tasmanian crayfish species: *Astacopsis franklinii*, which is a typical aquatic crayfish, and *Parastacoides tasmanicus*, which constructs burrows in swampland and routinely encounters drought during the summer. *Parastacoides* possesses more terminal spines on the filaments of the podobranchiae, which are assumed to prevent a collapse of the gills when exposed in air. They would also reduce clogging of the respiratory surface in the muddy water at the bottom of burrows. As crabs become increasingly terrestrial, there is a tendency for the gill surface area to become reduced (since it is a liability in terms of water loss) and be replaced by some accessory exchange surface such as a lung (see below, Greenaway 1984). It would appear that terrestrial crayfish adopt a different strategy, since *Parastacoides* has a significantly greater gill surface area than its aquatic counterpart. This is achieved by means of an increase in the number of gill filaments and a larger proportion of longer filaments. There are actually fewer gills (18 versus 21). The absence of pleurobranchiae increases the branchial space, which may facilitate the flow of particulate water and reduce clogging. It would also allow a larger volume of water to be retained within the branchial chamber when the animal is aerially exposed, which again would delay collapse. Despite a larger respiratory surface area, under normoxic conditions, *Parastacoides* has lower rates of O_2 utilization. Despite the presence of spines, there is considerable collapsing of the gills in air, which reduces the available surface area. Under these circumstances, the larger relative surface area coupled with the lower metabolic rate will allow it to tolerate aerial exposure better.

8.2 Air-Breathing

There have also been extensive studies on the respiratory adaptations of land crabs of FW origin, especially the species *Holthuisana* (Greenaway and Taylor 1976; Taylor and Greenaway 1979, 1984) and *Pseudothelphusa* (Innes and Taylor 1986; Innes et al. 1987). Both have similar life-styles in that they lead a fossorial existence (quiescence akin to estivation) during the dry season when water is not available (unlike marine land crabs which have continued access to SW). They then emerge from burrows to become active in the wet season. The evolution of crabs from aquatic to terrestrial environments has been correlated with a number of morphological modifications, such as a reduction in the gill surface area and evolution of a "lung" for aerial gas exchange, and physiological adjustments, such as a decrease in ventilation, regulation of the acid-base balance, and adjustments in

O_2 transport properties of the hemolymph. All of these areas have been recently reviewed by McMahon and Burggren (1988).

8.2.1 Morphology

The evolution towards air-breathing is generally associated with a reduction in the surface area of the gills (Greenaway 1984). As was seen above with terrestrial crayfish, gills in land crabs possess surface protruberances to prevent a collapse in air (Taylor and Greenaway 1979). In addition, the respiratory lamellae are strengthened with chitin, although this generally does not significantly increase the diffusion distance. The posterior gills, which in crabs are more important in ion regulation, are usually not reduced in land crabs, suggesting that their role continues.

Coupled with the reduction in gill area, the branchial chamber becomes enlarged, and its epithelial lining functions in gas exchange. The typical land crab lung epitomized in *Holthuisana* appears to function as a "cutaneous" aerial gas exchange organ, i.e., a flat, well-vascularized epithelium with chitin reduced to achieve a thin blood/air diffusion barrier of less than 1 μm (Greenaway and Taylor 1976; Taylor and Greenaway 1979, 1984). The lung surface area (SA) has been estimated as 10% of the gill SA in this crab.

Recently, it was discovered that one family of FW land crabs, the Pseudothelphusidae, possess an anastomosing, invaginated lung which is unique in terms of its structural complexity and its flow-through ventilation. Together these combine to make gas exchange perfusion, rather than diffusion, limited (Innes et al. 1987). The Trinidad mountain crab, *Pseudothelphusa garmani*, possesses a pair of lungs consisting of a system of primary and secondary anastomosing airways that are invaginations of the chitinized branchial chamber epithelium. These extend into a large hemolymph sinus situated in the anterior dorsolateral lining of each branchial chamber. A similarity between the exchange surface of this lung and that of birds and mammals has been suggested. The hemolymph-air diffusion distance is similar to that of the simple cutaneous lungs of other land crabs (around 0.4 μm). Based on ultrastructural studies, the lung acts exclusively in gas exchange; ion transport characteristics are retained by the reduced gills. The lungs appear to be perfused in parallel with the branchial chamber walls and the gills.

The morphological and physiological adaptations of the circulatory system in land crabs have been recently reviewed by Burggren and McMahon (1988b). The vasculature of the gills and lining of the branchial chamber exhibit certain refinements, although they follow the general pattern of aquatic species. The gill vasculature is virtually identical to that of aquatic species (Taylor and Greenaway 1979); however, the afferent and efferent supply to the branchiostegal membranes lining the branchial chamber is elaborated greatly. The afferent circulation varies substantially between families. For example, in *Holthuisana* the eye sinus serves as the main afferent supply. Branchial exchange occurs in the lacunar spaces arising from these afferent vessels; the diffusion distance is typically an order of magnitude lower than at the gills (for example 0.5 μm at the branchiostegal lung of *Pseudothelphusa garmani*, compared with 5 μm at the gill; El Haj et al. 1986). The hemolymph is gathered into large vessels that break up repeatedly into lacunae, constituting two portal systems with a total of three sets of lacunae in series. Anatomically, venous hemolymph draining any body tissue can perfuse either the gills or the branchial chamber, although there is some evidence for selective perfusion.

8.2.2 Physiology

When land crabs are immersed in water, the scaphognathites effectively pump water through the branchial chambers much as in aquatic species. When *Holthuisana* is respiring in water (Greenaway et al. 1983a), the weight-specific $\dot{M}O_2$ and $\dot{V}w$ are lower than predicted for primarily aquatic crabs, although the percentage extraction is higher. $\dot{M}O_2$ was not regulated at low ambient PO_2 but could be increased threefold during exercise. The postbranchial PO_2 was comparatively low; however, due to a high affinity, the respiratory pigment was saturated, and a substantial venous reserve was maintained. The $\dot{M}O_2$ in water, while low compared with other aquatic species, was higher than in air. This would appear to be correlated with active foraging while immersed. These crabs tend to remain relatively inactive in air.

During air-breathing in this species (Greenaway et al. 1983b), scaphognathite activity ceases, and the lungs are tidally ventilated by lateral oscillations of the thoracic walls. The neural mechanisms for switching between these oscillatory events is unknown but may involve mechanoreceptive and chemoreceptive information. Evidence for the latter includes observations that air-breathing animals exposed to extreme hypoxia will exhibit resumed scaphognathite activity (Greenaway et al. 1983b). Air enters and leaves via the prebranchial apertures and Milne-Edwards openings; the scaphognathites, which are generally used for ventilation in air or water in most land crabs, are inoperative. At rest, $\dot{V}w$ was low, suggesting a high diffusional component. Because air contains 40 times more O_2 than water, air flow through the branchial chambers is generally considerably lower than water flow for an equivalent O_2 uptake. Ventilation increased during hypercapnia (which would appear to be the primary stimulus for ventilation), hypoxia, and disturbance. As outlined above, the respiratory performance in air versus water seems to reflect O_2 requirements in the two media rather than the capacity of the respiratory system to function in each medium. Rates of CO_2 production have not been measured in any of the FW land crabs. There is evidence from marine land crabs that the gills may continue to be the primary exchange surface. However, there are many controversial aspects to the existing research on CO_2 excretion, such as whether or not the low rates reflect deposition of respiratory CO_2 in the carapace, and readers are urged to consult McMahon and Burggren (1988) for the most up-to-date review of this literature.

During settled air-breathing, the lungs and branchial chambers of *Pseudothelphusa* are ventilated by slow, intermittent, pressure/volume changes in the branchial chambers and not by scaphognathite action. Aerial ventilation appears to be unidirectional, which ensures that the branchial chamber PO_2 is above 140 torr and the postpulmonary hemolymph PO_2 is correspondingly high. This "bellows" system of ventilation is analogous to the air-sac system of birds.

As far as the control of ventilation is concerned, the FW land crabs appear to be transitional between aquatic species, which are sensitive to O_2 depletion, and aerial species, which respond to elevated CO_2 levels. *Holthuisana* has an increased CO_2 sensitivity during aerial ventilation, while retaining some sensitivity to O_2 during water-breathing (Greenaway et al. 1983a,b).

The slow, intermittent ventilation also serves to reduce the evaporative water loss (Innes and Taylor 1986); a similar pattern is seen in the crab *Holthuisana* (Greenaway and Taylor 1976). There is some suggestion that desiccating environments may have provided the selection pressure for the evolution of a complex lung with a high diffusing capacity and low ventilation requirement. The reduced $\dot{M}O_2$ during quiescence allows *Holthuisana* to maintain a comparatively high aerobic scope. A lung function analysis reveals that the lung of *Pseudothelphusa* is limited by convective conductance in the blood and not by diffusive conductance over the exchange surface, as in the cutaneous lung of other land crabs. The postpulmonary PCO_2 is low, suggesting that the crab can blow off CO_2 without resorting to an aquatic branchial route for CO_2 excretion.

The cardiovascular physiology in FW land crabs has to be largely inferred from morphological studies since hemodynamic measurements have only been directly made on marine species. The cardiac output is generally lower in land crabs, allegedly due to the increased hemolymph O_2 carrying capacity. As with reduced ventilation, the advantage of reduced perfusion would be to limit the evaporative water loss. A study by Taylor and Greenaway (1984), using the distribution of radiolabelled microspheres, provided evidence that the FW land crab *Holthuisana* can regulate the distribution of the peripheral circulation. While water-breathing, 80% of the $\dot{V}b$ is directed to the gills, whereas the reverse is true in air (i.e., 80% sent to the lungs). The switchover is incomplete, and the mechanism employed is unknown. It may involve active regulation or inactive regulation as a result of gill collapse in air. The latter would increase the branchial resistance and thereby redirect hemolymph to the lung.

8.3 Respiratory Pigments

The regulation of HCy O_2 affinity during aerial exposure of the crayfish *Austropotamobius pallipes* was investigated in a series of three papers (Morris et al. 1986a,b 1987). In the first, the interactive effects of Ca and lactate were investigated in vitro. The O_2 affinity of dialyzed hemolymph is increased by both increased lactate and Ca levels. Calcium would appear to be more effective on a molar basis in enhancing the O_2 affinity. However, the potentiating effects of these two modulators were found to be interdependent, i.e., elevating the concentration of one reduced the effect of the other. The increase in affinity between nondialyzed and dialyzed hemolymph could not be explained in terms of Ca and lactate alone, suggesting that there is at least one other unidentified factor involved. This initial study suggested that, during aerial exposure, the reduction in O_2 affinity one might expect as a result of the Bohr shift may not occur; instead the effects of Ca, lactate, and other factors may increase the affinity.

In a subsequent study (Morris et al. 1986b), the in vivo changes in O_2 affinity were investigated. In the opinion of those authors, the recovery of CaO_2 and CvO_2, which occurred after 24 h in air, could not be explained by the modest Bohr effect, as originally proposed by Taylor and Wheatly (1981). Based on their in vitro analysis, Morris et al. (1986b) predicted that the O_2 affinity of dialyzed hemolymph would increase during aerial exposure, initially due to an increase in lactate and subsequently due to an increase in Ca. The alkalosis induced by acid-base compensation would result further in a leftward shift in the dissociation curve. Large changes in the hemolymph O_2 content can be achieved by small changes in affinity, because HCy already has a low P_{50}.

In the final paper in the series, Morris et al. (1987) investigated further the interdependence between the potentiating effects of lactate and Ca on HCy affinity. They measured nine Ca binding sites per molecule of HCy and concluded that the interdependence was due to an interaction in the binding of the two ions, such that an increase in lactate of 3 mmol would be required to reduce by 1 the number of Ca ions bound to HCy when the Ca concentration was 17 m*M*. Since the two ions are differently charged, it is difficult to imagine competition at the same site. It is more likely that the observed effects result from conformational changes. The observation that elevating the concentration of one ion reduces the effect of the other may be adaptive during aerial exposure of the crayfish. If the effects of Ca and lactate were simply additive, then the HCy O_2 affinity might increase to such an extent that O_2 extraction by the tissues becomes impaired.

The O_2 binding properties of HCy in land crabs has been reviewed recently by McMahon and Burggren (1988), although relatively few studies have been conducted on FW species. There is a general tendency for the oxygen capacity to be greater in terrestrial decapods, which may allow for more O_2 to be picked up and transported. This in turn could allow for reduced branchial perfusion and, more importantly, reduced evaporative water loss. There has been no consensus as to whether adaptive changes in O_2 affinity occur with the transition to a terrestrial existence. Confusion stems from the fact that the animals vary in their activity levels, habitat, and the extent to which they depend on aquatic/aerial gas exchange. The general feeling is that we need a more complete understanding of the biology of these animals before factors influencing the evolution of their hemocyanins can be interpreted.

Recently, some research has been done on O_2 and CO_2 carriage by the HCy of the FW land crab *Holthuisana transversa* (Morris et al. 1988). Essentially, this study concludes that the *Holthuisana* HCy is insensitive to all modulating influences except temperature. The Bohr effect was small at all temperatures between 15 and 35 °C. There was no significant change in HCy concentration, O_2 affinity, or cooperativity with variation in Ca or L-lactate (which are generally potent modulators of the O_2 affinity of many hemocyanins). No specific effect of CO_2 on the O_2 affinity was found. It may be that the large changes in PCO_2 when *Holthuisana* switches from air to water, or during dehydration, make a CO_2-sensitive HCy maladaptive. A significant Haldane effect could not be demonstrated, which is not surprising in view of the low pH sensitivity. These authors conclude that a successful adaptation to a terrestrial environment may require a HCy insensitive to modulators normally considered important in aquatic species and that morphological, ventilatory, and cardiovascular responses are increasingly important in the regulation of hemolymph gas transport.

The temperature fluctuations in aerial environments are greater than in water. As well as having direct effects on HCy O_2 binding, they also may have indirect effects due to the concentration of circulating modulators caused by evaporative water loss at elevated temperatures. These questions were investigated by Morris et al. (1988). Within the temperature range 15–35 °C, increasing temperature decreased the O_2 affinity ($\Delta H = -54\,kJ\,mol^{-1}$) and reduced the amount of CO_2 that could be carried. Changes resulting from dehydration did not compromise O_2 transport. The sensitivity of O_2 binding to temperature appears to be higher than for other crustaceans.

8.4 Ion Balance

Desiccation is one of the major threats of terrestrial life. A negative water balance results primarily from evaporative loss across the respiratory surface. Once again, the physiological adaptations will be discussed for primarily aquatic FW species and then for those that routinely experience aerial exposure.

The effects of aerial exposure on the distribution of body water and ions have been investigated in the predominantly FW crayfish *Austropotamobius pallipes* (Taylor et al. 1987). As stated earlier, this species can generally only survive in air for around 3 days before succumbing. Crayfish exposed to water-saturated air (100% RH) did not lose mass, and the hemolymph levels of Na^+ and K^+ were maintained constant; however, on exposure to 70% RH, the crayfish progressively dehydrated at the rate of 0.38% body weight (BW) h^{-1} (0.5% body water h^{-1}) and died after 72 h, when 27% of initial mass had been lost. This fluid is lost by evaporation since the urine flow is minimal (Tyler-Jones and Taylor 1986). These values for degree of hydration and rate of water loss compare favorably with terrestrial crustaceans, suggesting that the FW crayfish is fairly well preadapted for dehydration. After a 10% reduction in mass, the hemolymph volume was reduced to 75% of its submerged value and was further reduced to 24% after 48 h. After 48 h the hemolymph osmolality had increased by 36%, with corresponding increases in the concentrations of K^+ and Cl^-, while that of Na^+ showed a smaller increase. Since crayfish can survive in hypersaline waters and seemingly tolerate the concentrations measured in the hemolymph at the time of death from aerial exposure, it would appear that the internal ion concentrations are not the ultimate cause of death. However, the reduction in circulating volume during severe dehydration no doubt limits the efficacy of the circulation, in terms of problems with increased viscosity and reduced O_2 delivery. All the same, the large volume/low pressure circulation in crustaceans may be suited to tolerate reduced volume better than the closed, high pressure system of vertebrates. The water content of whole muscle decreased slightly, indicating that the IC fluid volume (ICFV) is protected by preferentially losing water from the ECFV. The muscle Na^+ and K^+ concentrations increased, while that of Cl^- decreased. Proportional changes in the volume and ion concentrations of the different fluid compartments suggest that Na^+ was taken up from the ECF by the tissues, whereas Cl^- moved in the reverse direction, implicating the operation of ion mechanisms involved in the IC acid-base balance. There is also evidence to implicate the movement of ions and water between fluid compartments other than the hemolymph and muscles such as the foregut, which is known to be an ion storage organ in terrestrial decapods. The circulating Ca level doubled after 24 h in air; the increase was independent of the degree of dehydration and thus was thought to result from emersion and not dehydration per se. The increase was stoichiometrically equivalent to the observed increase in EC HCO_3^- and so may be due to mobilization of exoskeletal $CaCO_3$ to buffer the EC acidosis described earlier.

Although the gills are nonoperational while crayfish are exposed in air, the antennal gland could potentially be involved in ion and water balance at this time, and this was investigated by Tyler-Jones and Taylor (1986). When crayfish were exposed in air at 15 °C and 70% RH, the rate of urine flow decreased and ceased altogether after 24 h. On reimmersion after 48 h, the flow remained at zero for the initial 6 h, allowing the animals time to rehydrate, then gradually returned to normal levels. An increase in the inulin urine: hemolymph ratio during emersion suggests that the reduced flow is largely due to an increased reabsorption of water from the filtrate. Cessation of the urine flow (typically 8% BW day^{-1} while submerged) would constitute a doubling of the water loss rate (9% BW day^{-1}) in air if left unchecked. Therefore, it is an important factor in the overall water balance from 24 h onward. Urine, which is voided during the initial 24 h in air, will naturally drain into the branchial chambers, where it may reduce evaporative loss and may even be reabsorbed by the gills. Recycling of the urinary water has been proposed for aerial species such as the FW land crab *Holthuisana transversa* (Greenaway 1980). The urine concentrations of all ions except Ca^{2+} increased during aerial exposure, reflecting EC levels (see above). The urinary Ca^{2+} increased by only 24%, whereas the levels in the hemolymph more than doubled. This Ca may be selectively retained, perhaps for purposes of increasing the O_2 affinity of the HCy (see above).

While marine land crabs tend to have continued access to water, FW land crabs may routinely experience drought, so that desiccation is a more realistic environmental stress. The problems of ion and water balance in land crabs were recently reviewed by Greenaway (1988). Some research has been undertaken on FW land crabs, but there are no comparable studies on terrestrial crayfish. The primary problem of life out of water is dehydration. Freshwater species typically have a reduced water permeability (see above) which preadapts them for terrestrial living. However, loss of the contact with water means that the gills can no longer function in ion balance, and so other organs such as the antennal gland and gut become important in ion exchange. In addition to routine ion regulation, the molting cycle presents terrestrial species with additional ionic problems.

The rates of water loss have not been measured for many FW species, and those that do exist are questionable because of inconsistencies in the measurement technique. However, the lowest rates have been measured in highly terrestrial species, such as the arid zone crab *Holthuisana* (Greenaway and MacMillen 1978), compared with rates for marine or semi-terrestrial species. Even so, the rates are higher than comparably sized insects, indicating that water loss may have ultimately limited the extensive radiation of this subphylum into terrestrial environments. This is believed to be due to a lack of cuticular lipid in crustaceans. The rate of loss will reflect the temperature of the body surface, vapor pressure of the air, and rate of air movement across the body surface. While water loss will occur across the entire integument, the gills/lungs are the primary sites of loss, being thin and moist and having a large surface area (SA) by physiological design. The contribution of respiratory loss to the whole-animal water loss has not been accurately assessed. Another major site of water loss is urine, although the rates of urine flow are extremely low in terrestrial FW crabs. For example, *Holthuisana* produces urine at rates of 0.5 ml 100 $g^{-1}day^{-1}$ (Greenaway 1980) compared with 7.2 ml 100 $g^{-1}day^{-1}$ in crayfish (Wheatly and Toop 1989). There is also indirect evidence from potamonid crabs of extrarenal sites of water excretion (Shaw 1959a; Greenaway 1989), possibly in the midgut ceca or gills, but these have not been investigated further.

Freshwater land crabs tend to have higher body water contents and are able to tolerate a greater maximum water loss than their marine counterparts (Dandy and Ewer 1961; Lutz 1969; Greenaway and MacMillen 1978; Greenaway 1980; summarized in Table 7.3 of Greenaway 1988). Both of these attributes can be viewed as adaptations for living in desiccating environments. These crabs will generally immerse themselves in water if it is available and may burrow to ground water. Many marine species drink by spooning water into the mouth (or equally likely the gills) using the chelae or absorb moisture into the branchial chambers from the substrate using hydrophilic setae on their walking legs in conjunction with subambient pressures generated in the branchial chamber. Freshwater land crabs, however, do not use the latter mechanism because the heavy clays in which they burrow demand an extreme force for water removal, and the crabs do not possess the appropriate ventral setae (Greenaway and MacMillen 1978). Other water sources available to this crab are metabolic water from large fat deposits and the nightly condensation of water within the burrow, generated by diurnal temperature fluctuations. There is evidence that they exploit the latter by nocturnal vertical movements within the burrow.

Burrowing is a behavioral adaptation to limit water loss by creating a water-rich microenvironment. These crabs also select moist habitats and minimize activity at times of day or under climatic conditions which favor water loss. The FW land crabs from arid or monsoonal areas often experience long periods of drought and may spend extended periods (up to several years) in their burrows.

In terms of physiological adaptations for ion balance, FW land crabs are hyperosmotic in dilute water. The circulating electrolyte concentrations are lower than in their marine counterparts [total osmolality of around 500 mOsmol kg^{-1} in *Holthuisana transversa* (Greenaway and MacMillen 1978) and *Sudanonautes africanus* (Lutz 1969)]. Although they produce isosmotic urine, it does not constitute a major avenue for ion loss since the flow rate is so low (Harris and Micallef 1971; Greenaway 1980). The sodium regulation characteristics are typical of FW-adapted species, with low rates of diffusive loss and high affinity uptake mechanisms (Greenaway 1981). The major ionic problems encountered by FW land crabs are during dehydration. *Holthuisana* tolerates a loss of body water and associated hemolymph concentration without ostensibly regulating (Greenaway and MacMillen 1978). However, *Sudanonautes africanus* regulates the circulating Na^+ level while initially losing water from the hemolymph; with more extreme desiccation, water loss occurs from the tissue fluid and the hemolymph Na^+ content rises (Lutz 1969). Often these crabs do not feed during dry periods, and so they have no source for replacing ions lost via the urine. Their renal physiology has not been studied under terrestrial conditions. However, *Holthuisana* does release urine into the branchial chamber in water, and this would provide an opportunity for branchial ion reabsorption if it occurred on land.

At ecdysis, as outlined above, aquatic crustaceans increase their body volume by water absorption and then face a period of intense Ca^{2+} uptake to mineralize the new cuticle. Freshwater land crabs such as *Holthuisana* typically molt in water; however, they reabsorb 65% of the Ca^{2+}, Mg^{2+}, and phosphate from the preceding intermolt and store it in the hemolymph in the form of microspherules (0.25 µm) composed of amorphous calcium carbonate and phosphate with a small amount of magnesium (Sparkes and Greenaway 1984). This compares with 10–20% Ca storage in the majority of aquatic species, such as the FW crayfish. The hemolymph total Ca value was 150 times the intermolt value, but the Ca^{2+} concentration decreased from 4.5 to 3.4 m*M*. These microspherules undoubtedly increase the hemolymph viscosity and thereby the energy required for circulation. The site of production of the spherules is unknown. Following ecdysis, recalcification occurs extremely rapidly, and the spherules disappear within 12 h. Crabs obtain the remainder of their Ca requirement from their food and water; postmolt crabs exhibited a high affinity Ca^{2+} uptake mechanism (K_m = 0.105 mmol l^{-1}), which is very similar to that of the crayfish (0.13 mmol l^{-1}). Due to the large amount of stored Ca, they are less dependent on external water sources in the event that the available water dries up or is only present in small volumes. They do not appear to eat their exuviae in the field. Since these land crabs molt in FW, they presumably encounter similar ionic problems to those outlined earlier for crayfish, namely hemolymph dilution and the resulting uptake of Na^+ and Cl^-. However, these aspects have not been investigated.

9 Freshwater Challenges Attributable to Man

FW organisms are much more susceptible to exposure to toxicants of anthropogenic origin than marine species, because firstly, these are generally released into a more confined volume of water, and secondly, FW itself is less capable of buffering chemical insult due to its low background levels of transformation products. The more common problems arise from synthetic organic pesticides, industrial chemicals, petroleum hydrocarbons, and metal salts. Another major problem is acid rain, which originates when atmospheric acids are formed from sulfur oxides and nitrogen oxides emitted from factories and vehicle exhausts. Classically, toxicity has been evaluated as an all-or-none phenomenon as far as whether the animal survives the exposure. A more recent approach has been to assess sublethal exposure to ascertain the extent to which the organisms may be compromised physiologically. There have been some studies on the physiological effects of aquatic toxicants on FW crustaceans.

9.1 Heavy Metals

Heavy metal pollution orginates from the industry surrounding outcropping metalliferous lodes. These metals are absorbed into aquatic organisms passively (as with other ions, the permeability will be affected by external Ca^{2+}), via carrier systems for other ions, or may enter in the food (Bryan 1976). Others (e.g., manganese) may be incorporated into the cuticle during the molting cycle. Some of these are metals which are required in trace amounts for routine biochemistry; others are nonessential ions. The excretion and regulation of certain ions have been demonstrated. For example, the FW crayfish excretes zinc across the gut, while other metals are excreted across the gills (Bryan 1966, 1967, 1968). Temporary protection is afforded by transformation and storage. For example, the crayfish *Austropotamobius* can tolerate 30 times the usual blood concentration of zinc by binding it to blood proteins. The excess is then absorbed within a period of 2 days by the hepatopancreas (Bryan 1967), which is also the storage depot for excess copper and iron (Ogura 1959). One potential detoxification pathway, especially for copper, is binding into metallothioneins, which are nonenzymatic cytosolic proteins of low molecular weight with high concentrations of the essential metals, copper and zinc, as well as of the nonessential metals, cadmium, gold, mercury, and silver (Rainbow 1988). In decapods, there is good evidence that metallothioneins play a role in the normal physiological processes involving copper, which is an integral part of the respiratory pigment HCy. Metallothioneins have a high rate of turnover in cells, and lysosomal breakdown products were reported recently in the hepatopancreas cells of the crayfish (Chassard-Bouchard 1981). Calcium-rich granules (calcium carbonate or phosphate) also can bind, in an apparently detoxified form, a wide range of metals, including aluminum, silver, barium, cobalt, iron, manganese, lead, tin, strontium, and zinc (Rainbow 1988).

Since the early work by Bryan there have been few additional studies on FW crustaceans, although investigations on marine crustaceans and FW fish have delineated the effect of metals on their morphology, osmoregulation, and respiration. Generally, metal ions can structurally impair the branchial epithelium and reduce the gas diffusing capacity. They also may disrupt ion regulation by affecting active ion transport (e.g., ATPases and branchial permeability). In some species, an increased tolerance can be acquired by previous exposure to sublethal concentrations.

9.2 Acid Rain

Decapod crustaceans are very sensitive to acidification (Leivestad et al. 1976). France (1984, 1987a,b) has conducted intensive studies on the ecological effects of acid exposure on FW crayfish involving tolerance in the early life history stages, reproductive failure, bioaccumulation of heavy metals, and increased susceptibility to parasitic infections. Chronic lethality tests revealed that 2-week-old hatchlings were one-seventh as tolerant to acid stress as 2-month-old juveniles, which, in turn, were one-twentieth as tolerant as 2-year-old adults (France 1984), suggesting that acidification of a lake to below pH 5.5 could result in eventual population extinction as a result of mortality of the young. The mode of acid toxicity for crayfish in soft water is believed to be similar to that for fish, namely, failure to maintain internal ion homeostasis (Shaw 1960a; Morgan and McMahon 1982; see below). The age-specific mortality is probably a function of the general body size (i.e., small organisms have a greater SA to volume ratio, allowing for a faster rate of ion loss); it also may be related to physiological maturation. However, since little is known of the ontogeny of physiology in the crayfish, this is pure conjecture. As outlined above, the postmolt stages are more susceptible to acid stress (Malley 1980; Zanotto and Wheatly 1990, 1993a,b). Since the early stages molt more frequently, this exacerbates an already precarious situation.

Reproductive failure contributes to the crayfish population loss from FW lakes (France 1987b). The most significant effect of lake acidification on oviposition was the failure of eggs to attach to the pleopods due to incomplete cuticular hardening of the glair-cement forming the egg capsule and the stalk (a phemomenon similar to calcification problems after ecdysis). In addition, there was an increased egg mortality in acidified lakes. While this study only addressed egg viability prior to hatching, there was a suggestion that similar problems would prevent the hatchlings from attaching to the maternal pleopods by an elastic telson thread originating from the egg membrane that previously enclosed the embryo. The total recruitment of new individuals thus would be reduced greatly.

As outlined above (Sect. 6.3), inhibition of the postmolt calcification in acid waters has ramifying effects on crustaceans. France (1987a) showed that the decreased carapace rigidity in acid-exposed crayfish was associated with a reduced carapace dry weight and Ca content. Furthermore, the geochemistry of trace metals in FW now is known widely to be altered by acidification, such that the concentration increases. The tissue concentrations of Mn and Hg were both higher in crayfish from acidified lakes. In the case of Mn, the uptake is not only due to surface adsorption but to chemical competition with Ca and substitution into the structural layers of the cuticle.

Several studies have addressed the physiological effects of sublethal acid exposure in adult intermolt crayfish. Morgan and McMahon (1982) examined the effects of 4 days' exposure of *Procambarus clarki* to H_2SO_4 (Ca = 1 m*M*). They concluded that adult crayfish are remarkably acid tolerant (LC_{50} of pH 2.6), especially when compared with fish (LC_{50} generally around pH 4), possibly because the cuticle acts as a more effective barrier to H^+ influx. Obviously, the high tolerance of intermolt adults would be offset at the population level by the increased sensitivity of juveniles and postmolt animals. At sublethal pH (3.8), they observed a substantial hemolymph acidosis and a significant decrease in Na^+, although Cl^-, K^+, and Mg^{2+} levels were unaffected. The metabolic acidosis, which was greater than in corresponding studies on fish, may have originated internally from lactate accumulation or more likely externally via H^+ influx from the environment. The reduced circulating Na^+ may be due to an inhibition of Na^+ uptake and increased Na^+ efflux and could present severe ionic problems over longer exposures. During the first 24 h, there was a decrease in PCO_2, possibly reflecting an initial hyperventilation. The hemolymph Ca^{2+} also increased significantly, which they interpreted as dissolution of exoskeletal $CaCO_3$ to buffer the acidosis. In the long term, this would contribute to a reduced exoskeletal rigidity and would compound the existing problem of postmolt calcification in acid water (Malley 1980). Interestingly, crayfish were completely able to restore the ion and acid-base balance on return to neutral pH water after 4 days in acid, suggesting that no permanent morphological damage had been done to the branchial epithelium.

Wood and Rogano (1986) reassessed the physiological problems of acid toxicity in adult intermolt crayfish (*Orconectes*) using soft water (i.e., low Ca = 0.2 m*M*). In the wild, acidification is exclusively a soft water problem, and the fish literature has already shown that the Ca concentration has a critical modifying effect on toxicity. Five days' exposure to pH 4 (H_2SO_4) caused a severe EC metabolic acidosis and moderate reductions in the Na^+ and Cl^- content; however, the changes in these electrolytes were quantitatively not as large as the changes in the acid-base balance. The acidosis was due to an influx of H^+ from the environment. While the H^+ influx remained high, the accumulation of acidic equivalents in the ECF progressively slowed, suggesting that H^+ was being buffered in the ICF or the exoskeleton. Losses of Na^+ and Cl^- to water originated in the ECF; there was little evidence for exchange with the ICF or carapace. There was no lactate accumulation, and so presumably the O_2 delivery was unaffected. Contrary to Morgan and McMahon (1982), this study reported an initial elevation in $PaCO_2$ which contributed to the acidosis and may have resulted from hypoventilation or a thickened diffusion barrier at the gills. The hemolymph Ca^{2+} level rose and was lost to the environment; this Ca^{2+} could have originated either from the ICF or from the carapace buffering. Similar net effluxes of K^+ indicated that acidic equivalents, as suspected, penetrated the IC compartment. Unidirectional flux analysis (see Sect. 5.1.3) revealed that a negative Na^+ and Cl^- balance resulted in each case from the partial inhibition of influx; effluxes remained the same. All effects could be reversed during 5 days' recovery at pH 7.5.

A study on the combined effects of hypoxia (30% air saturation) and acid exposure (pH 4) on the hemolymph O_2 transport in intermolt adult crayfish was undertaken by Jarvenpaa et al. (1983). In neutral hypoxic water, the most pronounced change was an increase in the HCy O_2 affinity. In acid normoxic water, there was essentially no change in the transport characteristics. However, in acid hypoxic water, there was a decrease in the binding affinity and O_2 carrying capacity, believed to be due to the EC acidosis and wash-out of electrolytes. Therefore, it would appear that crayfish are unable to make the necessary compensations to hypoxia if simultaneously exposed to acid water.

It would be incorrect to assume that acid exposure is encountered only by FW crustaceans in areas polluted by man, since there are many examples in which it occurs in the wild. For example, the burrow water of the Tasmanian crayfish, *Parastacoides*, is characterised by a high organic content, with a pH ranging from 3.7–5.6, and O_2 tensions are commonly as low as 14 torr (Swain et al. 1987).

9.3 Nonmetal Pollutants

The major effects of nonmetal pollutants including pesticides (such as chlorinated hydrocarbons), industrial chemicals (such as polychlorinated biphenyls), and petroleum on marine crustaceans were reviewed recently by Fingerman (1985). Most of this research involves classic toxicology or effects on molting (inability to shed exoskeleton, reduced exoskeletal rigidity) and development (susceptibility of different larval stages), which are really special events in the life of an individual. There is very little information on routine physiological processes, such as respiration or ion regulation, and FW crustaceans have not been used extensively as test invertebrates even though they are a food item in parts of the world (crayfish are eaten in southern USA and in Scandinavia) and serve as food items for commercial fish. Again, the studies which exist (Muncy and Oliver 1963) focus on toxicity. This would appear to be a fertile area for future research.

References

Abele LG (1982) Biogeography. In: Abele LG (ed) The biology of crustacea, vol 1. Academic Press, New York, pp 241–304

Ahearn GA, Clay LP (1989) Kinetic analysis of electrogenic $2Na^+$-$1H^+$ antiport in crustacean hepatopancreas. Am J Physiol 257:R484–R493

Bergmiller E, Bielawski J (1970) Role of the gills in osmotic regulation in the crayfish *Astacus leptodactylus*. Comp Biochem Physiol 37:83–91

Bishop JA (1967) The zoogeography of the Australian freshwater decapod Crustacea. In: Weatherly AH (ed) Australian inland waters and their fauna. Australian National Univ Press, Canberra, pp 107–122

Blatchford JG (1971) Hemodynamics of *Carcinus maenas* (L.). Comp Biochem Physiol A 39:109–202

Bock F (1925) Die Respirationsorgane von *Potamobius astacus* Leach. (*Astacus fluviatilis* Fabr.). Ein Beitrag zur Morphologie der Decapoden. Z Wiss Zool 124:51–117

Born JW (1968) Osmoregulatory capacities of two Caridean shrimps, *Syncaris pacifica* (Atyidae) and *Palaemon macrodactylus* (Palaemonidae). Biol Bull 134:235–245

Bryan GW (1960) Sodium regulation in the crayfish *Astacus fluviatilis*. I. The normal animal. J Exp Biol 37:83–99

Bryan GW (1966) The metabolism of Zn and ^{65}Zn in crabs, lobsters and freshwater crayfish. In: Aberg B, Hungate FP (eds) Radioecological concentration processes. Pergamon Press, Oxford, pp 1005–1016

Bryan GW (1967) Zinc regulation in the freshwater crayfish (including some comparative copper analyses). J Exp Biol 46:281–296

Bryan GW (1968) Concentrations of zinc and copper in the tissues of decapod crustaceans. J Mar Biol Assoc U K 48:303–321

Bryan GW (1976) Some aspects of heavy metal tolerance in aquatic organisms. In: Lockwood APM (ed) Effects of pollutants on aquatic organisms. Cambridge University Press, Cambridge, pp 7–34

Burggren WW, McMahon BR (1988a) Biology of the land crabs: an introduction. In: Burggren WW, McMahon BR (eds) Biology of the land crabs. Cambridge University Press, New York, pp 1–5

Burggren WW, McMahon BR (1988b) Circulation. In: Burggren WW, McMahon BR (eds) Biology of the land crabs. Cambridge University Press, New York, pp 298–332

Burggren WW, McMahon BR, Costerton JW (1974) Branchial water and blood-flow patterns and the structure of the gill of the crayfish *Procambarus clarkii*. Can J Zool 52:1511–1518

Burtin B, Massabuau JC (1988) Switch from metabolic to ventilatory compensation of extracellular pH in crayfish. J Exp Biol 137:411–421

Burtin B, Massabuau JC, Dejours P (1986) Ventilatory regulation of extracellular pH in crayfish exposed to changes in water titration alkalinity and NaCl concentration. Respir Physiol 65:235–243

Cameron JN (1986) Acid-base equilibria in invertebrates. In: Heisler N (ed) Acid-base regulation in animals. Elsevier, New York, pp 357–394

Cameron JN (1989a) Acid-base homeostasis: past and present perspectives. Physiol Zool 62:845–865

Cameron JN (1989b) Post-moult calcification in the blue crab *Callinectes sapidus*: timing and mechanism. J Exp Biol 143:285–304

Cameron JN (1989c) The respiratory physiology of animals. Oxford University Press, New York

Cameron JN, Batterton CV (1978) Antennal gland function in the freshwater blue crab, *Callinectes sapidus*: water, electrolyte, acid-base and ammonia excretion. J Comp Physiol 123:143–148

Cameron JN, Mangum CP (1983) Environmental adaptations of the respiratory system: ventilation, circulation and oxygen transport. In: Vernberg FJ, Vernberg WB (eds) The biology of crustacea, vol 8. Academic Press, New York, pp 43–63

Chaisemartin C (1964) Importance des gastroliths dans l'économie du calcium chez *Astacus pallipes* Lereboullet. Bilan calcique de l'exuviation. Vie Milieu 15:457–474

Chassard-Bouchard C (1981) Rôle des lysosomes dans le phénomène de concentration du cadmium. Microanalyse par spectrographie des rayons X. C R Hebd Séances Acad Sci Paris Sér 3 293:261–265

Copeland DE, Fitzjarrell AT (1968) The salt absorbing cells in the gills of the blue crab (*Callinectes sapidus* Rathbun) with notes on modified mitochondria. Z Zellforsch 92:1–22

Cornell J (1976) Aspects of salt and water balance in two osmoconforming crabs, *Libinia emarginata* and *Pugettia producta* (Brachyura: Majidae). PhD Thesis, University of California, Berkeley (University Microfilms, Ann Arbor, Michigan)

Dandy JWT, Ewer DW (1961) The water economy of three species of the amphibious crab, *Potamon*. Trans R Soc S Afr 363:137–162

Dejours P, Armand J (1982) Variations de l'équilibre acid-base de l'hémolymphe d'ecrevisse en fonction des changements de certains propriétés physicochimiques de l'eau ambiante. C R Acad Sci Paris 295:509–512

Dejours P, Armand J (1983) Acid-base balance of crayfish hemolymph: effects of simultaneous changes of ambient temperature and water oxygenation. J Comp Physiol B 149:463–468
Dejours P, Beekenkamp H (1978) L'équilibre acid-base de l'hémolymphe au cours de la mue chez l'ecrevisse. C R Acad Sci Paris 286:1895–1898
Dejours P, Truchot JP (1988) Respiration of the emerged shore crab at variable ambient oxygenation. J Comp Physiol 158:387–391
Dejours P, Armand J, Beekenkamp H (1982) The effect of ambient chloride concentration changes on branchial chloride-bicarbonate exchanges and hemolymph acid-base balance of crayfish. Respir Physiol 48:375–386
Dickson GW, Franz R (1980) Respiration rates, ATP turnover and adenylate energy charge in excised gills of surface and cave crayfish. Comp Biochem Physiol A 65:375–380
Dickson JS, Dillaman RM (1985) Distribution and ultrastructure of osmoregulation and respiratory filaments in the gills of the crayfish. Am Zool 24:214
Ehrenfeld J (1974) Aspects of ionic transport mechanisms in crayfish *Astacus leptodactylus*. J Exp Biol 61:57–70
El Haj AJ, Innes AJ, Taylor EW (1986) Ultrastructure of the pulmonary, cutaneous and branchial gas exchange organs of the Trinidad mountain crab. J Physiol 373:84P
Fingerman SW (1985) Non-metal environmental pollutants and growth. In: Wenner AD (ed) Crustacean issues 3. Factors in adult growth. Balkema, Rotterdam, pp 219–234
Fisher JM (1972) Fine structural observations on the gill filaments of the fresh-water crayfish *Astacus pallipes* (Lereboullet). Tissue Cell 4:287–299
Flik G, Vanrijs JH, Wendelaar Bonga SE (1985) Evidence for high affinity Ca^{2+}-ATPase activity and ATP-driven Ca^{2+}-transport in membrane preparations of the gill epithelium of the cichlid fish *Oreochromis mossambicus*. J Exp Biol 119:335–347
France RL (1984) Comparative tolerance to low pH of three life stages of the crayfish *Orconectes virilis*. Can J Zool 62:2360–2363
France RL (1987a) Calcium and trace metal composition of crayfish *Orconectes virilis* in relation to experimental lake acidification. Can J Fish Aquat Sci 44:107–113
France RL (1987b) Reproductive impairment of the crayfish *Orconectes virilis* in response to acidification of lake 223. Can J Fish Aquat Sci 44:97–106
Gaillard S, Malan A (1983) Intracellular pH regulation in response to ambient hyperoxia or hypercapnia in the crayfish. Mol Physiol 4:231–243
Gaillard S, Malan A (1985) Intracellular pH-temperature relationship in a water breather, the crayfish. Mol Physiol 7:1–16
Gaillard S, Rodeau JL (1987) Na^+/H^+ exchange in crayfish neurons: dependence on extracellular sodium and pH. J Comp Physiol 157:435–444
Galler S, Moser H (1986) The ionic mechanism of intracellular pH regulation in crayfish muscle fibers. J Physiol (Lond) 374:137–151
Gannon AT, DeMarco VG, Morris T, Wheatly MG (1990) Metabolism and available oxygen for cave-dwelling crayfish. Am Zool 30:110A
Greenaway P (1970) Sodium regulation in freshwater mollusc *Limnaea stagnalis* (L) (Gastropoda: Pulmonata). J Exp Biol 53:147–163
Greenaway P (1972) Calcium regulation in the freshwater crayfish *Austropotamobius pallipes* (Lereboullet). I. Calcium balance in the intermoult animal. J Exp Biol 57:417–487
Greenaway P (1974a) Total body calcium and haemolymph calcium concentrations in the crayfish *Austropotamobius pallipes* (Lereboullet). J Exp Biol 61:19–26
Greenaway P (1974b) Calcium balance at the premoult stage of the freshwater crayfish *Austropotamobius pallipes* (Lereboullet). J Exp Biol 61:27–34
Greenaway P (1974c) Calcium balance at the postmoult stage of the freshwater crayfish *Austropotamobius pallipes* (Lereboullet). J Exp Biol 61:35–45
Greenaway P (1979) Fresh water invertebrates. In: Maloiy GMO (ed) Comparative physiology of osmoregulation in animals. Academic Press, London, pp 117–173
Greenaway P (1980) Water balance and urine production in the Australian arid-zone crab *Holthuisana transversa*. J Exp Biol 87:237–246

Greenaway P (1981) Sodium regulation in the freshwater/land crab *Holthuisana transversa*. J Comp Physiol B142:451–456
Greenaway P (1984) The relative importance of the gills and lungs in the gas exchange of amphibious crabs of the genus *Holthuisana*. Aust J Zool 32:1–6
Greenaway P (1985) Calcium balance and moulting in the crustacea. Biol Rev 60:425–454
Greenaway P (1988) Ion and water balance. In: Burggren WW, McMahon BR (eds) Biology of the land crabs. Cambridge University Press, New York, pp 211–248
Greenaway P, MacMillen RE (1978) Salt and water balance in the terrestrial phase of the inland crab *Holthuisana (Austrothelphusa) transversa* Martens (Parathelphusoidea: Sundathelphusidae). Physiol Zool 51:217–229
Greenaway P, Taylor HH (1976) Aerial gas exchange in Australian arid-zone crab *Parathelphusa transversa* Von Martens. Nature (Lond) 262:711–713
Greenaway P, Bonaventura J, Taylor HH (1983a) Aquatic gas exchange in the freshwater/land crab *Holthuisana transversa*. J Exp Biol 103:225–236
Greenaway P, Taylor HH, Bonaventura J (1983b) Aerial gas exchange in Australian freshwater/land crabs of the genus *Holthuisana*. J Exp Biol 103:237–251
Harris RR (1975) Urine production rate and urinary sodium loss in the fresh water crab *Potamon edulis*. J Comp Physiol 96:143–153
Harris RR, Micallef H (1971) Osmotic and ionic regulation in *Potamon edulis*, a fresh water crab from Malta. Comp Biochem Physiol A 38:769–776
Hassall CH (1979) Respiratory physiology of the crayfish *Procambarus clarki*. MSc Thesis, University of Calgary, Calgary, Alberta Canada.
Henry RP (1984) The role of carbonic anhydrase in blood ion and acid-base regulation. Am Zool 24:241–251
Henry RP, Cameron JN (1982) The distribution and partial characterization of carbonic anhydrase in selected aquatic and terrestrial decapod crustaceans. J Exp Zool 221:309–321
Henry RP, Wheatly MG (1992) Interaction of respiration, ion regulation, and acid-base balance in marine crabs. Am Zool 32:407–416
Hughes GM, Knights B, Scammel CA (1969) The distribution of PO_2 and hydrostatic pressure changes within the branchial chambers of the shore crab *Carcinus maenas*. J Exp Biol 51:203–220
Huxley TH (1879) The crayfish. Kegan, Paul, Trench, London
Innes AJ, Taylor EW (1986) The evolution of air-breathing in crustaceans: a functional analysis of branchial, cutaneous, and pulmonary gas exchange. Comp Biochem Physiol A 85:621–637
Innes AJ, Taylor EW, El Haj AJ (1987) Air-breathing in the Trinidad mountain crab: a quantum leap in the evolution of the invertebrate lung. Comp Biochem Physiol A 87:1–9
Jarvenpaa T, Nikinmaa M, Westman K, Soivio A (1983) Effects of hypoxia on the haemolymph of the freshwater crayfish, *Astacus astacus* L., in neutral and acid water during the intermolt period. In: Goldman CR (ed) Freshwater crayfish. V. Papers from the 5th Int Symp on Freshwater crayfish. AVI, Westport CT, pp 86–97
Jay D, Holdich DM (1981) The distribution of the crayfish, *Austropotamobius pallipes*, in British waters. Freshwater Biol 11:121–129
Kamemoto FI, Keister SM, Spalding AE (1962) Cholinesterase activities and sodium movement in the crayfish kidney. Comp Biochem Physiol 7:81–87
Kerley DE, Pritchard AW (1967) Osmotic regulation in the crayfish *Pacifastacus leniusculus* stepwise acclimated to dilutions of seawater. Comp Biochem Physiol 20:101–113
Kirschner LB (1979) Control mechanisms in crustaceans and fishes. In: Gilles R (ed) Mechanisms of osmoregulation in animals. Wiley Interscience, Chichester, pp 157–222
Kirschner LB, Greenwald L, Kerstetter TH (1973) Effect of amiloride on sodium transfer across body surface of freshwater animals. Am J Physiol 224:832–837
Larimer JL, Gold AH (1961) Responses of the crayfish, *Procambarus simulans*, to respiratory stress. Physiol Zool 134:167–176

Leivestad H, Hendrey G, Muniz IP, Snekvik E (1976) Effects of acid precipitation on freshwater organisms. In: Braekke FH (ed) Impact of acid precipitation on forest and freshwater ecosystems in Norway. SNSF Project FR6/76, Oslo, pp 87–111

Lowenstam HA, Weiner S (1989) On biomineralization. Oxford University Press, New York, pp 120–122

Lowery RS (1988) Growth, moulting and reproduction. In: Holdich DM, Lowery RS (eds) Freshwater crayfish: biology, management and exploitation. Croom Helm, London, pp 83–113

Lutz PL (1969) Salt and water balance in the West African fresh water/land crab *Sudanonautes africanus africanus* and the effects of desiccation. Comp Biochem Physiol 30:469–480

Malley DF (1980) Decreased survival and calcium uptake by the crayfish *Orconectes virilis* in low pH. Can J Fish Aquat Sci 37:364–372

Malley DF, Chang PSS (1985) Effects of aluminum and acid on calcium uptake by the crayfish *Orconectes virilis*. Arch Environ Contam Toxicol 14:739–747

Maluf NSR (1941) Secretion of inulin, xylose, and dyes and its bearing on the manner of urine formation by the kidney of the crayfish. Biol Bull 8:235–260

Mangum CP (1983) Oxygen transport in the blood. In: Mantel LH (ed) The biology of crustacea, vol 5. Academic Press, London, pp 373–429

Mangum CP (1985) Molting in the blue crab *Callinectes sapidus*: a collaborative study of intermediary metabolism, respiration and cardiovascular function, and ion transport. Preface. J Crustacean Biol 5:185–187

Mantel LH, Farmer LL (1983) Osmotic and ionic regulation. In: Mantel LH (ed) The biology of crustacea, vol 5. Academic Press, London, pp 53–161

Massabuau JC, Burtin B (1984) Regulation of oxygen consumption in the crayfish *Astacus leptodactylus* at different levels of oxygenation: role of peripheral O_2 chemoreception. J Comp Physiol 155:43–49

Massabuau JC, Burtin B (1985) Ventilatory CO_2 reflex response in hypoxic crayfish *Astacus leptodactylus* acclimated to 20 °C. J Comp Physiol 156:115–118

Massabuau JC, Eclancher B, Dejours P (1980) Ventilatory reflex response to hyperoxia in the crayfish, *Astacus pallipes*. J Comp Physiol 140:193–198

Maynard DM (1960) Circulation and heart function. In: Wolverkamp HP, Waterman TH (eds) The physiology of crustacea, vol I. Academic Press, New York, pp 161–226

McLaughlin PA (1983) Internal anatomy. In: Mantel LH (ed) The biology of crustacea, vol 5. Academic Press, London, pp 1–52

McMahon BR (1981) Oxygen uptake and acid-base balance during activity in decapod crustaceans. In: Herreid CF II, Fourtner CR (eds) Locomotion and energetics in arthropods. Plenum Press, New York, pp 299–335

McMahon BR, Burggren WW (1988) Respiration. In: Burggren WW, McMahon BR (eds) Biology of the land crabs. Cambridge University Press, New York, pp 249–297

McMahon BR, Hassall CD (1979) Ventilation and oxygen transport in resting and active crayfish acclimated to cool temperature. Physiologist 22(4):85

McMahon BR, Wilkens JL (1983) Ventilation, perfusion, and oxygen uptake. In: Mantel LH (ed) The biology of crustacea, vol 5. Academic Press, London, pp 289–372

McMahon BR, Wilkes PRH (1983) Emergence responses and aerial ventilation in normoxic and hypoxic crayfish *Orconectes rusticus*. Physiol Zool 56:133–141

McMahon BR, Burggren WW, Wilkens JL (1974) Respiratory responses to long-term hypoxia stress in the crayfish *Orconectes virilis*. J Exp Biol 60:195–206

McWhinnie MA (1962) Gastrolith growth and calcium shifts in the freshwater crayfish *Orconectes virilis*. Comp Biochem Physiol 7:1–14

Moody Jr WJ (1980) Appearance of calcium action potentials in crayfish slow muscle fibres under conditions of low intracellular pH. J Physiol (Lond) 302:335–346

Moody Jr WJ (1981) The ionic mechanism of intracellular pH regulation in crayfish neurones. J Physiol (Lond) 316:293–308

Morgan DO, McMahon BR (1982) Acid tolerance and effects of sublethal acid exposure on ionoregulation and acid-base status in two crayfish *Procambarus clarki* and *Orconectes rusticus*. J Exp Biol 97:241–252

Morris S, Tyler-Jones R, Taylor EW (1986a) The regulation of haemocyanin oxygen affinity during emersion of the crayfish *Austropotamobius pallipes*. I. An in vitro investigation of the interactive effects of calcium and L-lactate on oxygen affinity. J Exp Biol 121:315–326
Morris S, Tyler-Jones R, Bridges CR, Taylor EW (1986b) The regulation of haemocyanin oxygen affinity during emersion of the crayfish *Austropotamobius pallipes*. II. An investigation of in vivo changes in oxygen affinity. J Exp Biol 121:327–337
Morris S, Bridges CR, Grieshaber MK (1987) The regulation of haemocyanin oxygen affinity during emersion of the crayfish *Austropotamobius pallipes*. III. The dependence of Ca^{2+}-haemocyanin binding on the concentration of L-lactate. J Exp Biol 133:339–352
Morris S, Greenaway P, McMahon BR (1988) Oxygen and carbon dioxide transport by the haemocyanin of an amphibious crab, *Holthuisana transversa*. J Comp Physiol 157:873–882
Moshiri GA, Goldman CR, Godshalk GL, Mull DR (1970) The effect of variations in oxygen tension on certain aspects of respiratory metabolism in *Pacifastacus leniusculus* (Dana) (Crustacea: Decapoda). Physiol Zool 43:23–29
Muncy RJ, Oliver AD (1963) Toxicity of ten insecticides to the red crawfish *Procambarus clarki* (Girard). Trans Am Fish Soc 92:428–431
Mykles DL (1980) The mechanism of fluid absorption at ecdysis in the American lobster, *Homarus americanus*. J Exp Biol 84:89–101
Ogura K (1959) Midgut gland cells accumulating iron or copper in the crayfish *Procambarus clarkii*. Ann Zool Jpn 32:133–142
Ortmann AE (1902) The geographical distribution of freshwater decapods and its bearing upon ancient geography. Proc Am Philos Soc 41:267–400
Parry G (1957) Osmoregulation in some fresh water prawns. J Exp Biol 34:417–423
Parry G, Potts WTW (1965) Sodium balance in the fresh water prawn *Palaemonetes antennarius*. J Exp Biol 42:415–421
Pennak RW (1989) Fresh-water invertebrates of the United States. Wiley, New York, 628pp
Peterson DR, Loizzi RF (1974) Ultrastructure of the crayfish kidney coelomosac, labyrinth, nephridial canal. J Morphol 142:241–263
Rainbow PS (1988) The significance of trace metal concentrations in decapods. In: Fincham AA, Rainbow PS (eds) Aspects of decapod crustacean biology. Clarendon Press, Oxford, pp 291–313
Riegel JA (1963) Micropuncture studies of chloride concentration and osmotic pressure in the crayfish antennal gland. J Exp Biol 40:487–492
Riegel JA (1968) Analysis of the distribution of sodium, potassium and osmotic pressure in the urine of crayfishes. J Exp Biol 48:587–596
Riegel JA (1972) Comparative physiology of renal excretion. Oliver and Boyd, Edinburgh
Rodeau JL (1982) L'état acid-base intracellulaire: analyse Théorique appliquée à l'érythrocyte des Mammifères et étude expérimentale des cellules nerveuses et musculaires des Crustacés. Thèse Etat Sciences, L'université Louis Pasteur de Strasbourg
Roer R, Dillaman R (1984) The structure and calcification of the crustacean cuticle. Am Zool 24:893–909
Rutledge PS (1981) Effects of temperature acclimation on crayfish haemocyanin oxygen binding. Am J Physiol 240:R93–R98
Rutledge PS, Pritchard AW (1981) Scope for activity in the crayfish *Pacifastacus leniusculus*. Am J Physiol 240:R87–R92
Schram FR (1982) The fossil record and evolution of Crustacea. In: Abele LG (ed) The biology of Crustacea; systematics, the fossil record and biogeography, vol 1. Academic Press, New York, pp 93–147
Shaw J (1959a) Salt and water balance in the East African fresh-water crab, *Potamon niloticus* (M. Edw.). J Exp Biol 36:157–176
Shaw J (1959b) The absorption of sodium ions by the crayfish, *Astacus pallipes* Lereboullet. I. The effect of external and internal sodium concentration. J Exp Biol 36:126–144

Shaw J (1960a) The absorption of sodium ions by the crayfish *Astacus pallipes*. II. The effect of the external anion. J Exp Biol 37:534–547
Shaw J (1960b) The absorption of sodium ions by the crayfish *Astacus pallipes*. III. The effect of other cations in the external solution. J Exp Biol 37:557–572
Shaw J (1961) Sodium balance in *Eriocheir sinensis* (M. Edw.). The adaptation of the Crustacea to fresh water. J Exp Biol 38:153–162
Shaw J (1964) The control of salt balance in the Crustacea. Symp Soc Exp Biol 18:237–256
Shetlar RE, Towle DW (1989) Electrogenic sodium-proton exchange in membrane vesicles from crab (*Carcinus maenas*) gill. Am J Physiol 257:R924–R933
Short TM, Haswell MS (1979) Ionic and osmotic adjustments of the crayfish *Orconectes immunis* in response to dilute external concentration of sodium sulfate and choline chloride. Am Zool 19:906
Simkiss K, Wilbur KM (1989) Biomineralization. Academic Press, San Diego, pp 205–289
Sinha NP, Dejours P (1980) Ventilation and blood acid-base balance of the crayfish as functions of water oxygenation (40-1500 Torr). Comp Biochem Physiol A 65:427–432
Sparkes S, Greenaway P (1984) The haemolymph as a storage site for cuticular ions during premoult in the freshwater/land crab *Holthuisana transversa*. J Exp Biol 113:43–54
Sutcliffe DW (1975) Sodium uptake and loss in *Crangonyx pseudogracilis* (Amphipoda) and some other crustaceans. Comp Biochem Physiol A 52:255–257
Swain R, Marker PF, Richardson AMM (1987) Respiratory responses to hypoxia in stream-dwelling (*Astacopsis franklinii*) and burrowing (*Parastacoides tasmanicus*) parastacid crayfish. Comp Biochem Physiol A 87:813–817
Swain R, Marker PF, Richardson AMM (1988) Comparison of the gill morphology and branchial chambers in two fresh-water crayfishes from Tasmania: *Astacopsis franklinii* and *Parastacoides tasmanicus*. J Crustacean Biol 8:355–363
Taylor EW (1981) Some effects of temperature on respiration in decapodan crustaceans. J Therm Biol 6:239–248
Taylor EW (1982) Control and co-ordination of ventilation and circulation in crustaceans: responses to hypoxia and exercise. J Exp Biol 100:289–319
Taylor EW, Wheatly MG (1980) Ventilation, heart rate and respiratory gas exchange in the crayfish *Austropotamobius pallipes* (Lereboullet) submerged in normoxic water and following 3 h exposure in air at 15°C. J Comp Physiol 138:67–78
Taylor EW, Wheatly MG (1981) The effect of long-term aerial exposure on heart rate, ventilation, respiratory gas exchange and acid-base status in the crayfish *Austropotamobius pallipes*. J Exp Biol 92:109–124
Taylor EW, Tyler-Jones R, Wheatly MG (1987) The effects of aerial exposure on the distribution of body water and ions in the freshwater crayfish *Austropotamobius pallipes* (Lereboullet). J Exp Biol 128:307–322
Taylor HH, Greenaway P (1979) The structure of the gills and lungs of the arid-zone crab, *Holthuisana* (*Austrothelphusa*) *transversa* (Martens) (Sundathelphusidae: Brachyura) including observations on arterial vessels within the gills. J Zool (Lond) 189:359–384
Taylor HH, Greenaway P (1984) The role of the gills and branchiostegites in gas exchange in a bimodally breathing crab, *Holthuisana transversa*: evidence for a facultative change in the distribution of the respiratory circulation. J Exp Biol 111:103–122
Taylor HH, Taylor, EW (1986) Observations of valve-like structures and evidence for rectification of flow within the gill lamellae of the crab *Carcinus maenas* (Crustacea, Decapoda). Zoomorphology 106:1–11
Travis DF (1960) The deposition of skeletal structures in the Crustacea. I. The histology of the gastrolith in the crayfish, *Orconectes* (*Cambarus*) *virilis* Hagen-Decapoda. Biol Bull 118:137–149
Travis DF (1963) Structural features of mineralization from tissues to macromolecular levels of organization in the decapod crustacea. Ann N Y Acad Sci 109:177–245

Truchot JP (1983) Regulation of acid-base balance. In: Mantel LH (ed) The biology of crustacea, vol 5. Academic Press, London, pp 431–457

Truchot JP (1987) Comparative aspects of extracellular acid-base balance. Springer, Berlin Heidelberg New York, 248pp

Tyler-Jones R, Taylor EW (1986) Urine flow and the role of the antennal glands in water balance during aerial exposure in the crayfish *Austropotamobius pallipes* (Lereboullet). J Comp Physiol B 156:529–535

Tyler-Jones R, Taylor EW (1988) Analysis of haemolymph and muscle acid-base status during aerial exposure in the crayfish *Austropotamobius pallipes*. J Exp Biol 134:409–422

Vernberg FJ (1983) Respiratory adaptations. In: Vernberg FJ, Vernberg WB (eds) The biology of crustacea, vol 8. Academic Press, New York, pp 1–42

Walsh PJ, Milligan CL (1989) Coordination of metabolic and intracellular acid-base status: ionic regulation and metabolic consequences. Can J Zool 67:2994–3004

Wheatly MG (1985a) Free amino acid and inorganic ion regulation in the muscle and haemolymph of the blue crab *Callinectes sapidus* (Rathbun) in relation to the molting cycle. J Crustacean Biol 5:223–233

Wheatly MG (1985b) The role of the antennal gland in ion and acid-base regulation during hyposaline exposure of the Dungeness crab *Cancer magister* (Dana). J Comp Physiol B 155:445–454

Wheatly MG (1989) Physiological responses of the crayfish *Pacifastacus leniusculus* (Dana) to environmental hyperoxia. I. Extracellular acid-base and electrolyte status and transbranchial exchange. J Exp Biol 143:33–51

Wheatly MG (1990) Postmolt electrolyte regulation in crayfish: Ca budget, hemolymph ions and tissue Ca ATPase. Am Zool 30:63A

Wheatly MG (1993) An overview of electrolyte regulation in the freshwater crayfish throughout the molting cycle. In: Romaire RP (ed) Freshwater crayfish, vol 8. (in press)

Wheatly MG, Gannon AT (1993) The effect of external electrolytes on postmolt calcification in the freshwater crayfish *Procambarus clarkii* (Girard). In: Holdich DM (ed) Freshwater crayfish, vol 9. (in press)

Wheatly MG, Henry RP (1987) Branchial and antennal gland Na^+/K^+-dependent ATPase and carbonic anhydrase activity during salinity acclimation of the euryhaline crayfish *Pacifastacus leniusculus*. J Exp Biol 133:73–86

Wheatly MG, Ignaszewski LA (1990) Electrolyte and gas exchange during the molting cycle of a freshwater crayfish. J Exp Biol 151:469–483

Wheatly MG, Taylor EW (1981) The effect of progressive hypoxia on heart rate, ventilation, respiratory gas exchange and acid-base status in the crayfish *Austropotamobius pallipes*. J Exp Biol 92:125–141

Wheatly MG, Toop T (1989) Physiological responses of the crayfish *Pacifastacus leniusculus* (Dana) to environmental hyperoxia II. The role of the antennal gland. J Exp Biol 143:53–70

Wheatly MG, Toop T, Morrison RJ, Yow LC (1991) Physiological responses of the crayfish *Pacifastacus leniusculus* (Dana) to environmental hyperoxia. III. Intracellular acid-base balance. Physiol Zool 64:323–343

Wilkes PRH, McMahon BR (1982a) Effect of maintained hypoxic exposure on the crayfish *Orconectes rusticus*. I. Ventilatory, acid-base and cardiovascular adjustment. J Exp Biol 98:119–137

Wilkes PRH, McMahon BR (1982b) Effect of maintained hypoxic exposure on the crayfish *Orconectes rusticus*. II. Modulation of haemocyanin oxygen affinity. J Exp Biol 98:139–149

Willig A, Keller R (1973) Molting hormone content, cuticle growth and gastrolith growth in the molt cycle of the crayfish *Orconectes limosus*. J Comp Physiol 86: 377–388

Wood CM, Boutilier RG (1985) Osmoregulation, ionic exchange, blood chemistry, and nitrogenous waste excretion in the land crab *Cardisoma carnifex*: a field and laboratory study. Biol Bull 169:267–290

Wood CM, Rogano MS (1986) Physiological responses to acid stress in crayfish (*Orconectes*): haemolymph ions, acid-base status, and exchanges with the environment. Can J Fish Aquat Sci 43:1017–1026
Zanotto FP, Wheatly MG (1990) Postmolt calcification in crayfish as a function of ambient pH in normal and decarbonated fresh water. Physiologist 33(4):A37
Zanotto FP, Wheatly MG (1993a) The effect of pH on postmolt calcification and ion regulation in the freshwater crayfish (*Procambarus clarkii*). In: Romaire RP (ed) Freshwater crayfish, vol 8. (in press)
Zanotto FP, Wheatly MG (1993b) The effect of ambient pH on electrolyte regulation during postmoult in freshwater crayfish *Procambarus clarkii*. J Exp Biol (in press)

Chapter 4

Neurochemical Differences in Two Steroid-Sensitive Areas Mediating Reproductive Behaviors

Jacques Balthazart[1] and *Gregory F. Ball*[2]

Contents

1 Introduction

Extensive progress has been made in the past 20 years in the elucidation of the mechanism of gonadal steroid action in the brain underlying the activation of reproductive behavior (McEwen et al. 1987; Harlan 1988; Blaustein and Olster 1989; Pfaff 1989a). This progress has been greatly facilitated by the use of relatively few model systems that have been investigated in detail. One of the best studied hormone-dependent reproductive behaviors is the lordosis response in rodents which is required for successful female mating (Pfaff 1980; Pfaff and Schwartz-Giblin 1988). This has been a popular model system through which to understand hormone-behavior relationships since the 1930s (Beach 1948). With the identification of the ventromedial nucleus (VMN) of the hypothalamus as the primary site in the brain for the activa-

[1] Laboratory of General and Comparative Biochemistry, University of Liège, 17 place Delcour, B-4020 Liège, Belgium
[2] Department of Psychology, The Johns Hopkins University, Baltimore, MD 21218, USA

Advances in Comparative and Environmental Physiology, Vol. 15

tion by steroids of this behavior, a series of elegant biochemical investigations were initiated in several laboratories which have now identified a cascade of neurochemical steps initiated by estradiol-17β (E_2) and progesterone (P) that seem necessary for the natural occurrence of the behavior in the intact animal. This analysis is now being pursued at the molecular level, and specific mRNAs implicated in the behavioral response are in the process of being isolated and identified (McEwen and Pfaff 1985; McEwen et al. 1987; Pfaff and Schwartz-Giblin 1988; Pfaff 1989b; Simerly et al. 1989). It is now reasonable to expect that in the years to come there will be a coherent explanation of lordosis in the female rat at the molecular level.

Although this concentration on a few model systems has been successful, it is important to remember that significant differences in the biochemical consequences of steroid action may exist among different species and even within a species among different steroid-sensitive brain areas (McEwen et al. 1984). In different species, a homologous pattern of behavior appears to be activated by different steroid metabolites. For example, female receptivity can be activated by E_2 alone in some species (e.g., voles: Carter and Getz 1985), while in other species the sequential action of E_2 and P appears to be absolutely necessary (e.g., rats and guinea pigs: Feder 1984; McEwen et al. 1987). Within a species, different strains also exhibit varying steroid requirements for the activation of a given behavior (e.g., in mice: Luttge and Hall 1973; Luttge 1979). These differences may represent arbitrary and even insignificant variations in the physiological "solution" of the general problem of the hormonal activation of behavior or may just be artifacts, merely reflecting variations in the experimental methods employed. Alternatively, these differences may provide insight into a more general theory of how hormones act biochemically to activate behavior.

The steroid-sensitive nuclei underlying different reproductive behaviors, even within an individual animal, show a diversity of biochemical mechanisms in generating the steroid-dependent response (McEwen et al. 1984). Their comparison could reveal essential principles of brain functioning without the possible confounding effects that are inevitable when comparisons across species or strains are attempted. Thus, the experimental analysis of steroid actions on the neurochemistry in different brain regions which are implicated in the control of various behaviors is another strategy to utilize for understanding hormone action in the brain.

This comparative approach within the same species has been used in a series of studies on the Japanese quail. In this chapter, we would like to review and contrast the biochemical substrates underlying the activation by steroids of two reproductive behaviors: copulation and the male-typical vocalization, crowing. In the quail, both of these behaviors are activated by testosterone (T): they disappear quickly after castration and can be restored within days by T replacement therapy (Adkins and Pniewski 1978; Adkins-Regan 1983; Balthazart et al. 1983). The sexually dimorphic medial preoptic nucleus (POM) appears to be a major site of T action for the restoration of copulatory behavior (Balthazart and Surlemont 1990a,b). However, T implants in or near the POM never elicited crowing in the Japanese quail (Watson and Adkins-Regan 1989b). Rather, the mesencephalic nucleus intercollicularis (ICo) appears to be an important area mediating crowing and other vocalizations in the quail (Potash 1970; Armitage and Seller 1981). We therefore compared and contrasted the behavioral effects of steroids, the metabolism of steroids, and the occurrence of steroid receptors in these two areas. Additional attention is given to the modulation of the neurochemical systems by steroids in these two brain nuclei.

2 The Biochemistry of Testosterone Action in the Quail Brain

2.1 The Neuroanatomical Sites of Testosterone Action

Stereotaxic implants and electrophysiological studies have facilitated the localization of brain areas that can be considered the primary sites of T action in the activation of reproductive behavior. Cannulae filled with crystalline T activate copulatory behavior when implanted in the POM, but not in adjacent regions of the preoptic area (POA) or other steroid-sensitive areas of the hypothalamus (Fig. 1). These implants have no effect on crowing behavior (Watson and Adkins-Regan 1989b; Balthazart and Surlemont 1990a). The role of the POM in the activation of copulatory behavior is confirmed by experiments in which electrolytical lesions were placed in the POA (see Fig. 2). These lesions affected the activation by T of the masculine sexual behavior only if they destroyed a significant fraction of the POM. There is a highly significant correlation ($r > 0.90$) between the percentage of the nucleus which had been destroyed and the behavioral deficit, and the latter was not related to the absolute size of the lesion (Balthazart and Surlemont 1990a).

While no study in quail has demonstrated that stereotaxic implants of T in the ICo activate crowing, it is well documented that steroid-filled cannulae located in the ICo in ring doves specifically elicit the nest-coo vocalizations in males and females and the bow-coo vocalization in males only; these vocalizations are a part of the courtship ritual (Cohen 1981, 1983; Cohen and Cheng 1981, 1982). However, steroid implants in the ICo failed to elicit crowing in capons (Phillips and Barfield, 1977), though this may have been related to the testing conditions. In quail, ICo lesions have been demonstrated to decrease, although not totally suppress, a number of vocalizations including crowing (Seller 1981). In addition, the electrical simulation of ICo elicits several vocalizations including crowing (Armitage and Seller 1981). Therefore, this area participates in the production of the crowing vocalization, and steroid are quite likely to modulate the behavior by acting in this area as will be discussed below.

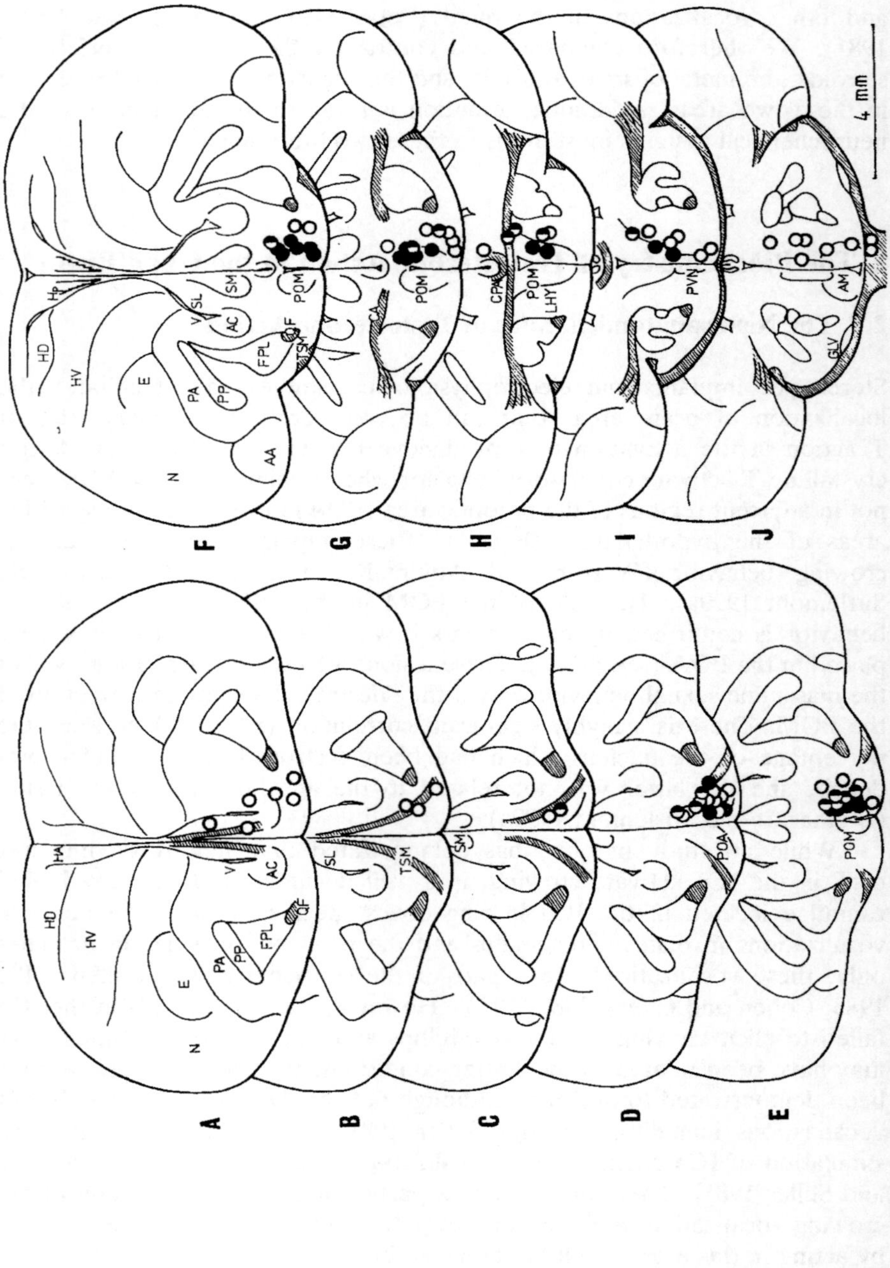
A
B
C
D
E
F
G
H
I
J
N
E
HV
HD
HA
Hp
PA
PP
FPL
QF
AC
V
SL
SM
TSM
POA
POM
AA
CA
CPA
LHY
PVN
AM
GLV
4 mm

2.2 Behavioral Effects of Testosterone and Its Metabolites

In the Japanese quail, T activates a number of reproductive behaviors such as courtship, copulatory sequence (neck grab, mount, and cloacal contact movement), and several vocalizations associated with reproduction, among which crowing is one of the most prominent (Adkins and Adler 1972; Adkins 1977; Balthazart et al. 1983). Some of these behaviors are elicitable after T treatment in gonadectomized males and females (e.g., crowing, although there are quantitative and perhaps qualitative differences between males and females), while others can only be activated in males and will seldom if ever be seen in females even after treatment with supraphysiological doses of T (e.g., cloacal contact movements) (Adkins and Adler 1972; Adkins 1975; Adkins and Pniewski 1978; Balthazart et al. 1983). Interestingly, copulation and crowing seem to be activated by T through its transformation to different metabolites (Adkins 1977; Adkins et al. 1980; Schumacher and Balthazart 1983). The aromatization of T into E_2 seems to be a critical step in the activation of copulatory behavior in that E_2 or estradiol benzoate (EB) treatment activates the behavior, inhibition of the aromatization process by specific compounds such as the aromatase inhibitor androstatrienedione (ATD) prevents the occurrence of T-induced but not E_2-induced behavior, and T-induced behavior can be blocked by antiestrogens such as tamoxifen and CI628 (see Balthazart 1989 for review). The nonaromatizable androgen, 5α-dihydrotestosterone (5α-DHT), alone has little or no effect on copulation but seems to synergize with E_2 in the activation of this behavior (Adkins and Pniewski 1978; Adkins et al. 1980; Balthazart et al. 1985). By contrast, crowing is effectively elicited in castrated quail by treatment with 5α-DHT alone (Adkins and Pniewski 1978). E_2 has no effect on this behavior, and in one study it was even reported that E_2 blocked the facilitation by 5α-DHT of this behavior in

←

Fig. 1. Transverse sections of the quail brain illustrating the locations of the unilateral T implants which did (*black circles*) or did not (*open circles*) restore masculine sexual behavior in castrated male quail. Sections **A** to **J** are arranged in a rostral to caudal order, and each is 180 μm apart. The center of the stereotaxic needles is indicated. Implant sites are coded according to the behavior shown by the birds as follows: ●, showed mount attempt (MA) regularly (at least 3 out of the last 4 tests); ◑, showed MA irregularly; ○, never showed MA. *AA* Nucleus archistriatalis; *AC* nucleus accumbens; *AM* nucleus anterior medialis hypothalami; *CA* commissura anterior; *CPA* commissura pallii; *E* ectostriatum; *FPL* fasciculus prosencephali lateralis; *GLV* nucleus geniculatus lateralis pars ventralis; *HA* hyperstriatum accessorium; *HD* hyperstriatum dorsale; *Hp* hippocampus; *HV* hyperstriatum ventrale; *LHY* nucleus lateralis hypothalami; *N* neostriatum; *PA* paleostriatum augmentatum; *POA* preoptic area; *POM* medial preoptic nucleus; *PP* paleostriatum primitivum; *PVN* nucleus periventricularis magnocellularis; *QF* tractus quintofrontalis; *SL* nucleus septalis lateralis; *SM* nucleus septalis medialis; *TSM* tractus septomescencephalicus; *V* ventricle. The neuroanatomical nomenclature is based on Baylé et al. (1974) and Panzica et al. (1987). (Balthazart and Surlemont 1990a)

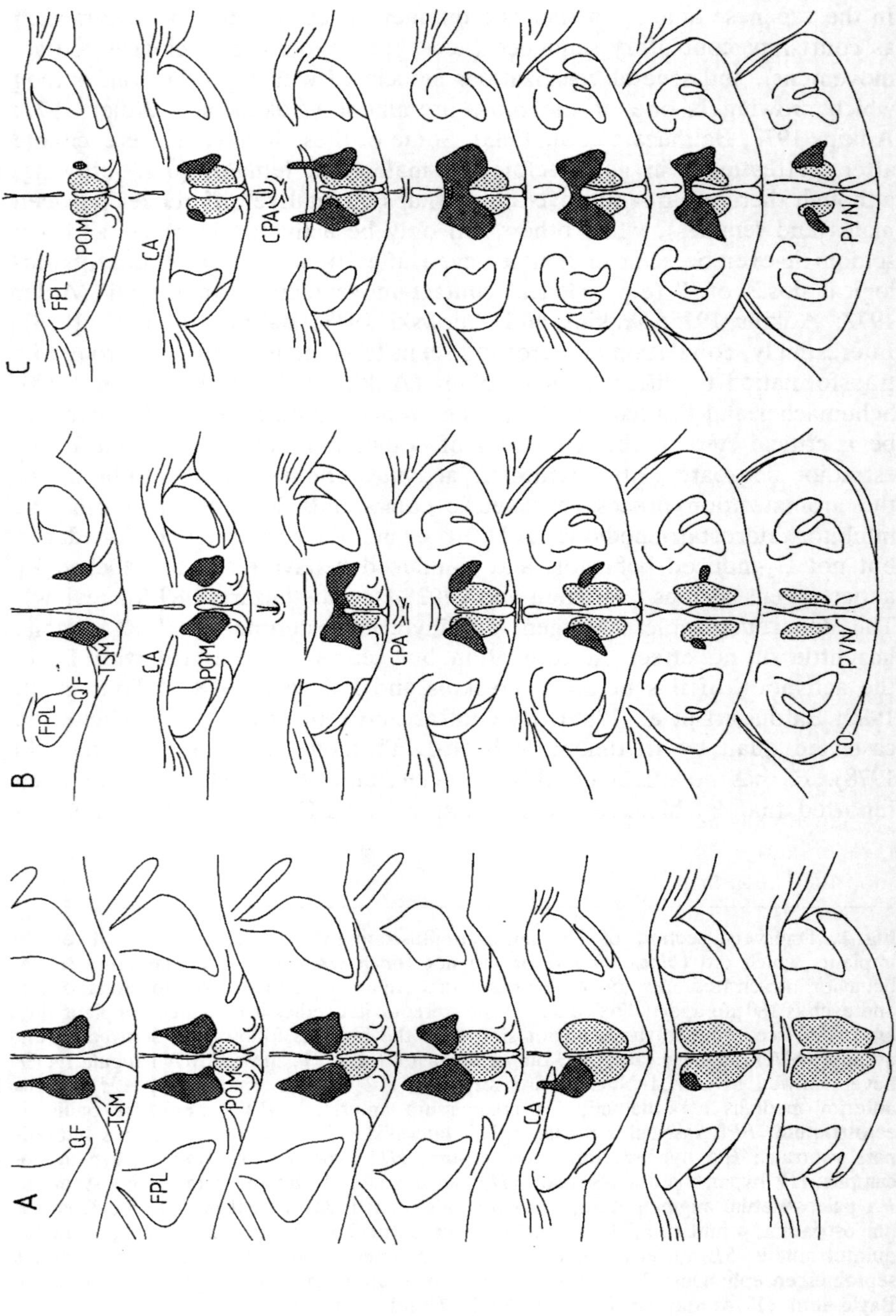
A
QF
TSM
FPL
POM
CA
B
FPL
QF
TSM
CA
POM
CPA
PVN
CO
C
FPL
POM
CA
CPA
PVN

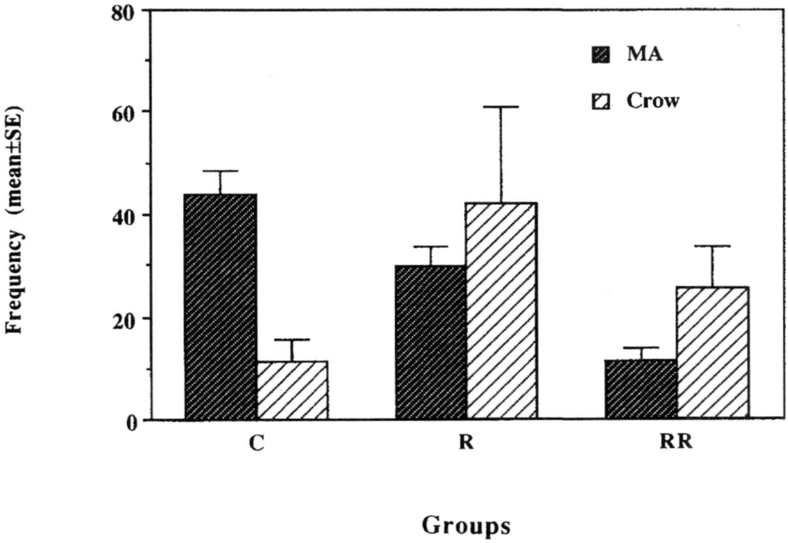

Fig. 3. Effects of unilateral (*R*) or bilateral (*RR*) implants of the aromatase inhibitor R76713 and of control implants filled with cholesterol (*C*) on the testosterone-induced reproductive behavior in castrated quail. The figure shows the total frequency of mount attempts (*MA; dense stippling*) and of crows (*light stippling*) observed during the behavior tests. MA was significantly affected by the treatments ($F_{2,42} = 19.62$, $P < 0.0001$), but no significant change in the crow frequency was observed ($F_{2,42} = 1.53$, NS). (Based on data from Balthazart et al. 1990a)

young chicks (Schumacher and Balthazart 1985). The study of the pharmacological blockade of T effects on crowing has generated some contradictory results, in that both antiandrogens and antiestrogens decrease the activation by T of this vocalization (Alexandre and Balthazart 1986). However, the potent aromatase inhibitor R76713 clearly blocked the T-induced copulatory behavior without having any effect on the crowing frequency (Balthazart et al. 1990a). The effects on copulation were apparently induced by an action in the POA as bilateral implants of R76713 in this region significantly inhibited the behavior (Fig. 3). Taken together, the available evidence strongly suggests that copulation is primarily E_2-dependent, while crowing is essentially dependent on androgen action in the brain.

←

Fig. 2A–C. Transversal reconstruction of a preoptic area lesion which did not inhibit sexual behavior **A** and of a lesion which partially **B** or totally **C** suppressed it. The behaviorally effective lesion overlaps to a large extent with the medial preoptic nucleus (POM). In each case, sections are organized in the rostral to caudal direction from the top to the bottom of the figure. *Dark area* lesion; *dotted area* POM. See legend of Fig. 1 for abbreviations. (Based on data from Balthazart and Surlemont 1990a)

2.3 Brain Metabolism of Testosterone

The hormonal specificity in the control of copulation and crowing is reflected at the level of the metabolism of T, which occurs in the brain areas that appear to underlie those behaviors. As stated above, specific regions of the quail brain are capable of transforming T into both E_2 and 5α-DHT by the action of two enzymes, aromatase and 5α-reductase, respectively. A sensitive radioenzymatic assay demonstrated that the POM in males contains very high levels of aromatase activity, while this enzyme is undetectable in ICo (Balthazart et al. 1990f). Recently, the anatomical localization of the aromatase has become possible with the application of immunocytochemical techniques. These studies have confirmed the findings of the enzyme assays in that large numbers of aromatase-immunoreactive cells have been revealed in the POM, while they appear to be totally absent in the ICo (Balthazart et al. 1990b). Conversely, 5α-reductase activity is much higher in the ICo than in the POM (Balthazart et al. 1990f).

The aromatase activity in the POM is modulated by T, the substrate for the enzyme. Castration markedly decreases the aromatase activity, and T treatment reinstates it to values comparable to those seen in intact, sexually mature animals (Schumacher and Balthazart 1986; Balthazart et al. 1990f). Similar changes in the number of aromatase-immunoreactive neurons in the POM are seen after castration and T replacement therapy (Balthazart et al. 1990b). This type of regulation is not observed for the 5α-reductase in the POM nor in the ICo (Balthazart et al. 1990f; Fig. 4). Thus, the enzymes that produce the behaviorally active metabolites are located in the appropriate sections of the brain which are thought to be a part of the neural substrate underlying the two different behaviors (i.e., estrogen-dependent copulation in the POM and androgen-dependent crowing in ICo). There are further correlations between the occurrence of the enzymes and the production of the behaviors. As mentioned above, copulation is sexually differentiated in quail (i.e., it can be activated by T in adult males but not in adult females; Adkins and Adler, 1972; Adkins 1975), and parallel to this, the aromatase activity is higher in males than in females (Schumacher and Balthazart 1986). In contrast, crowing can be elicited with T in males and females, and 5α-reductase shows no sexual dimorphism in the ICo (Balthazart et al. 1990f).

2.4 Steroid Receptors

Both the POM and ICo in the Japanese quail contain steroid receptors, as demonstrated recently by autoradiographic and immunohistochemical studies (Balthazart et al. 1989b; Watson and Adkins-Regan 1989a). The POM is characterized by a high density of E_2 receptors, as shown by autoradiography and immunohistochemistry. [^{3}H]T labeled a moderate

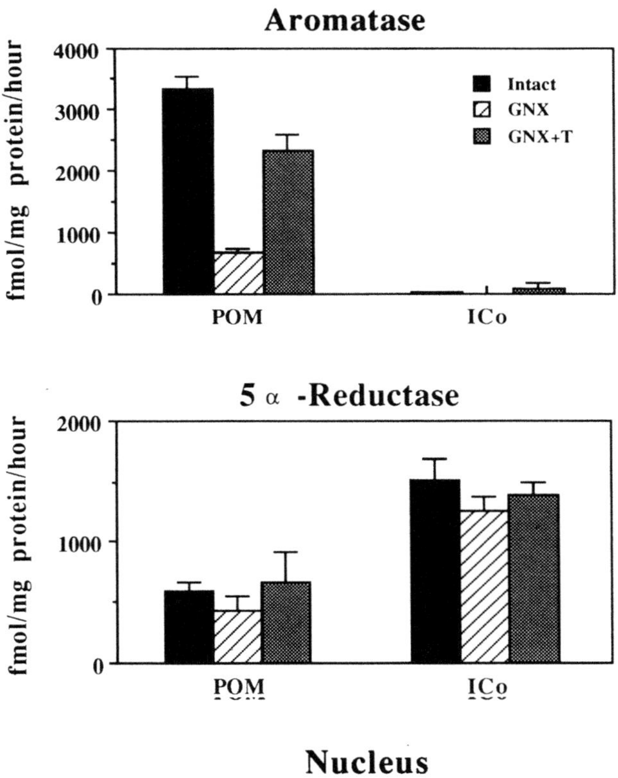

Fig. 4. Activity of aromatase and 5α-reductase in the medial preoptic nucleus (*POM*) and in the nucleus intercollicularis (*ICo*) of male Japanese quail that were either sexually mature with intact gonads (*Intact*) or had been gonadectomized (*GNX*) or gonadectomized and submitted to a replacement therapy with testosterone (*GNX + T*). Enzyme activities were measured by an in vitro radioenzyme product-formation assay as described in Schumacher and Balthazart (1986). Data are means ± standard errors of the amounts of metabolites produced. (Based on data from Balthazart et al. 1990f)

number of cells in the POM, but it is possible that the labeled hormone was converted to E_2 before binding to the E_2 receptors. This interpretation is supported by the fact that following the injection of [^{3}H] DHT, very few labeled cells were observed in the POM (Watson and Adkins-Regan 1989a). Therefore, it is possible that the POM contains very few or no androgen receptors. The ICo contains both E_2 and androgen receptors (Watson and Adkins-Regan 1989a; Balthazart et al. 1989b). A large number of labeled cells were found in the ICo following the injection of either [^{3}H] E_2 or [^{3}H] DHT, and the presence of E_2 receptors was confirmed by immunohistochemistry (Fig. 5).

In line with the behavioral and metabolic data reviewed above, it appears that E_2 receptors are present in a larger proportion of cells in the

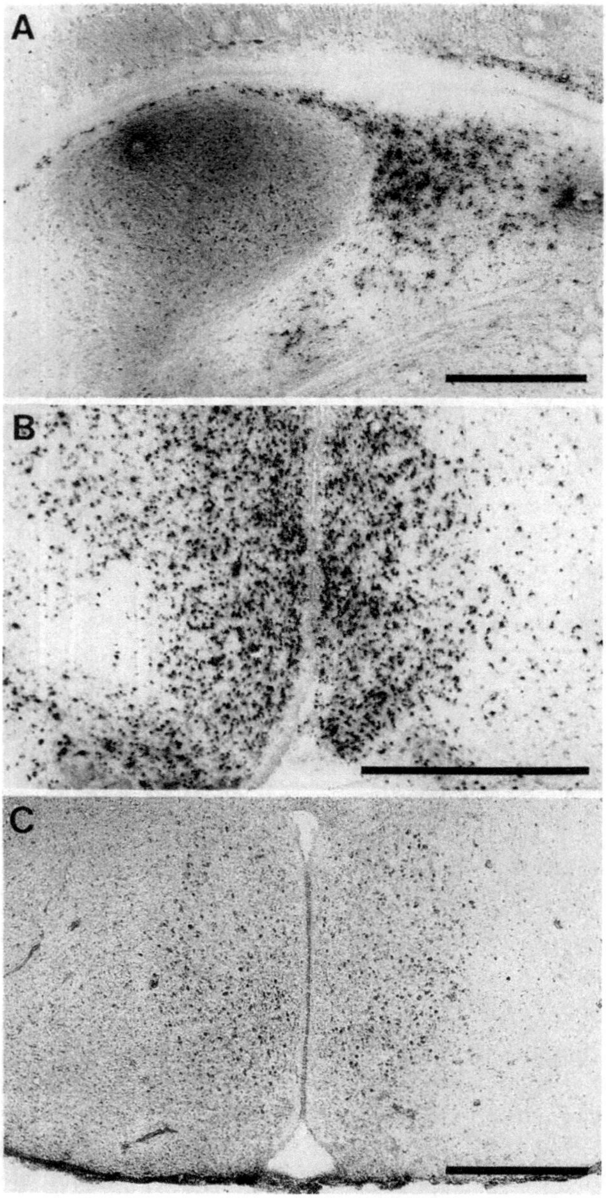

Fig. 5A–C. Microphotographs illustrating the distribution of immunoreactive estrogen receptors in the nucleus intercollicularis (*ICo*; **A**) and in the medial preoptic nucleus (*POM*; **B**) as well as the distribution of immunoreactive aromatase in the POM (**C**) of male Japanese quail. No aromatase-immunoreactive neurons are found in ICo. *Magnification bar* is 500 µm in each case (magnification is ×50 in **A**, **C** and ×80 in **B**). (Based on data from Balthazart et al. 1989a, 1990b, Ball et al. 1989b)

POM (approximately 90%) than in the ICo (approximately 50%; Balthazart et al. 1989b). There is an anatomical correlation between the presence of receptors and of T-metabolizing enzymes in these two brain nuclei: Aromatase activity is high in the POM where almost all cells contain E_2 receptors, while 5α-reductase activity is highest in the ICo where autoradiographic studies detected one of the highest levels of androgen receptors in the quail brain. However, there are two problems with these apparent correlations. The relationship between aromatase activity and the presence of the estrogen receptor could not be confirmed at the cellular level based on colocalization studies using immunocytochemistry (Fig. 6). Only 10% of the cells in the POM was found to contain immunoreactive aromatase, and less than 20% of these was found also to stain positively for the presence of immunoreactive estrogen receptors (Balthazart et al. 1990c,d).

It is also somewhat surprising to find that about one-half of the cells in the ICo contain E_2 receptors when the levels of aromatase are undetectable

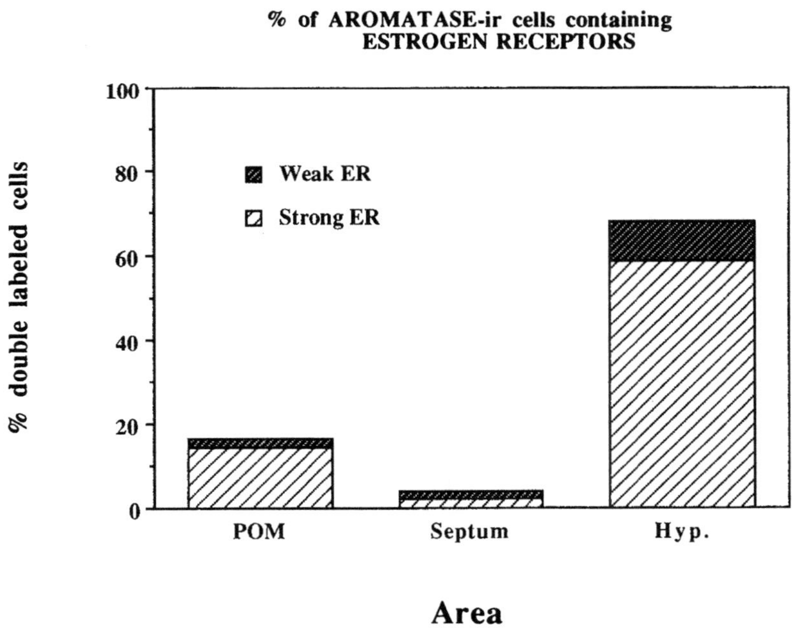

Fig. 6. Quantitative analysis of the colocalization of aromatase and estrogen receptors in the quail brain. Counts were made separately in 3 regions of the brain: the preoptic medial nucleus (*POM*), the septal region ventral to the nucleus accumbens and the bed nucleus of the stria terminalis (*Septum*), and the cluster of cells running from the dorsal part of the ventromedial nucleus of the hypothalamus to the nucleus inferioris hypothalami (*Hyp.*). Data are means of 4 males. A distinction between dark and light immunostaining of estrogen receptors (*ER*) has been made in the counts. This indicates that a small proportion only of the ER immunoreactive cells are at the limit of detection of the procedure. (Based on data from Balthazart et al. 1990d)

(Balthazart et al. 1990f). There are, of course, significant levels of circulating E_2 in both male and female quail (Balthazart et al. 1987; Watson et al. 1990). Estrogen receptors in ICo could play a role in mediating a behavior that is directly controlled by these estrogens secreted into the general circulation (e.g., other vocalizations; Guyomarc'h 1974).

3 Steroid-Neurotransmitter Interactions in the POM and ICo

One of the ways in which steroids influence behavior is by initiating a series of neurochemical changes in specific areas of the brain (McEwen et al. 1984, 1987; Dohanich et al. 1985; McEwen and Pfaff 1985). The investigations of hormonal mechanisms underlying reproductive behavior have progressed from the elucidation of hormone-sensitive sites to the study of the kind of biochemical changes that hormones induce in these areas. The most notable changes involve differences in neurotransmitter activity. Again, the study of this link between steroids and neurotransmitter functioning has progressed the most with investigations of the lordosis response in the laboratory rat (McEwen et al. 1987, 1988). In the quail, such studies have been initiated with the analysis of the catecholaminergic and cholinergic functioning in several steroid-sensitive areas in the brain. In this section, we describe the steroid effects in the POM and ICo on neurotransmitter receptors as measured by quantitative autoradiography and the effects on neurotransmitter levels and turnover measured in microdissected nuclei with high performance liquid chromatography (HPLC) and electrochemical detection (EC).

3.1 Monoamine Levels and Turnover and Cholinergic Activity

Norepinephrine, epinephrine, dopamine, and serotonin baseline levels have been measured in the POM and the ICo of quail (Balthazart and Ball 1989; J. Balthazart and P. Sante, unpubl. data). Levels of norepinephrine are 5–10 times higher in the POM than in the ICo. The other amines are also more concentrated in the POM than in the ICo, but the magnitude of the difference is usually not as substantial. This might indicate that in the POM, relative to the ICo, noradrenergic transmission plays a more important role.

Acetylcholinesterase appears to be a specific marker of the ICo in a number of avian species (Ryan and Arnold 1981; Vischer et al. 1982; Gardner and Harding 1990). It does not heavily mark the preoptic area. No such data are available at present for the Japanese quail.

The steady-state levels of norepinephrine in the ICo are higher in females than in males (Fig. 7). This sex difference is not affected by castration or T replacement therapy (data not shown). In contrast, in the POM females also have higher norepinephrine levels than in males, but in gonadectomized animals no such difference is apparent. The decline in the norepinephrine concentration observed following the intraperitoneal injection of α-methyl-*para*-tyrosine (αMPT), a tyrosine hydroxylase inhibitor, was lower in the POM than in the ICo (no decrease was observed in many cases). This might be taken as evidence that the turnover rate of norepinephrine is much lower in the POM than in the ICo. However, it is also possible that the extremely high concentration of norepinephrine in the POM is associated with a high concentration of tyrosine hydroxylase, whose activity was therefore not completely blocked by the dose of inhibitor utilized in this study.

The steady-state levels of dopamine were appreciably higher in the POM of males than in females. In addition, αMPT caused a dramatic decline in the concentration of this amine in males, but it had no clear effect in females. This difference was observed in both gonadally intact and gonadectomized animals, suggesting that it represents a sexually dimorphic character that was organized by steroids early in life and is not sensitive to the activational effects of these hormones. Such a sex difference in dopamine activity was not seen in the ICo. With the exception of the sex difference in the POM norepinephrine concentration referred to previously, sex steroids had no apparent effect on the steady-state levels or on the turnover rates of norepinephrine and dopamine in both nuclei.

3.2 Neurotransmitter Receptors

Quantitative autoradiography has been employed to characterize the distribution of the α_1 and α_2 adrenergic receptor subtypes and the muscarinic cholinergic receptor in both the POM and ICo (Ball et al. 1989a,b, 1990; Balthazart and Ball 1989; Balthazart et al. 1989a). The α_1 and α_2 adrenergic receptors were labeled, respectively, with the antagonist [^{3}H] prazosin (PRZ) and the agonist [^{3}H] *para*-aminoclonidine (PAC), while the muscarinic receptors were revealed by the binding of [^{3}H] *n*-methylscopolamine (NMS), an antagonist. All of these receptors have been shown to be modulated by steroids in areas of the rodent brain that mediate steroid-dependent reproductive processes (Dohanich et al. 1985; Johnson et al. 1985, 1988). α_1 Receptors were undetectable in the POM (Balthazart et al. 1989a). The density of α_2-adrenergic receptors in the POM is among the highest attained in any area of the brain. Interestingly, the pattern of [^{3}H] PAC binding corresponds precisely to the shape of the POM nucleus as defined by Nissl studies; this is true throughout the rostral to caudal extent of the nucleus (Ball et al. 1989a). There is no sex difference in receptor density apparent when running a one-point assay that samples the binding at a single rostral-caudal level. The binding does not seem to be affected by castration. When the nucleus is sampled throughout the rostral to caudal extent using a saturation analysis within an individual, there is also no detectable effect of

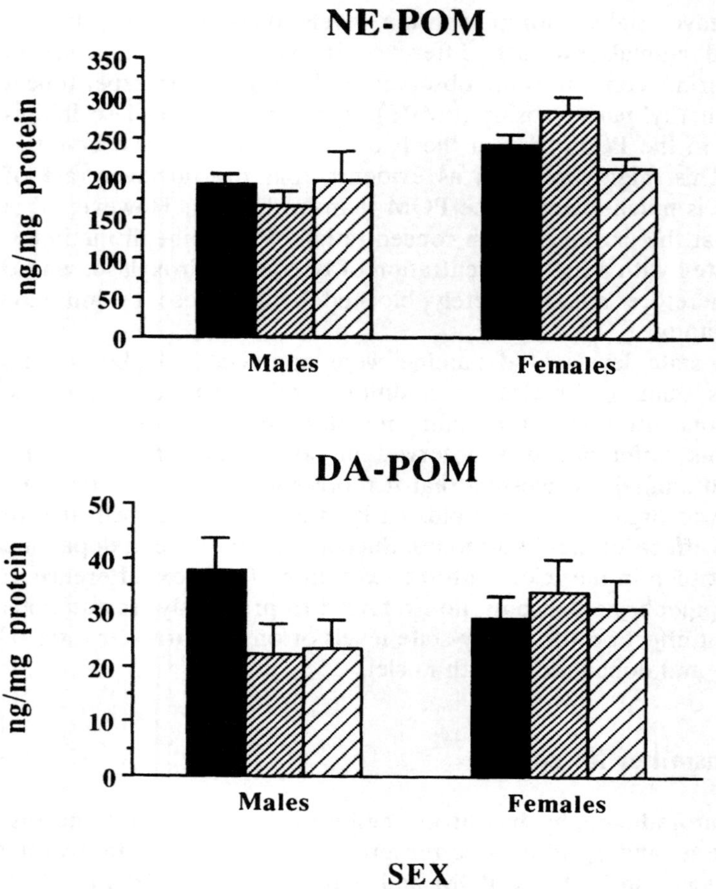

Fig. 7. Effects of α-methyl-*para*-tyrosine (αMPT), a tyrosine hydroxylase inhibitor, on the levels of norepinephrine (*NE*) and dopamine (*DA*) in the medial preoptic nucleus (*POM*) and in the nucleus intercollicularis (*ICo*) of male Japanese quail. Brain samples were collected before (*t0*) or 90 (*t90*) or 180 (*t180*) min after the injection of the inhibitor. Nuclei were dissected by the Palkovits punch technique and assayed by high performance liquid chromatography coupled with electrochemical detection. Data are means ± standard errors. (J. Balthazart and P. Sante, unpubl. data)

castration or T treatment on either the maximum number of binding sites or the apparent dissociation constant (Balthazart et al. 1991; Foidart et al. 1990).

In contrast to the pattern described above for [^{3}H] PAC, muscarinic cholinergic ([^{3}H] NMS) binding defines the POM by the absence of a significant amount of binding (Ball et al. 1990). Binding in the POM is close to the background level and is much lower than that apparent in the adjacent lateral POA. These low levels of binding do not appear to be

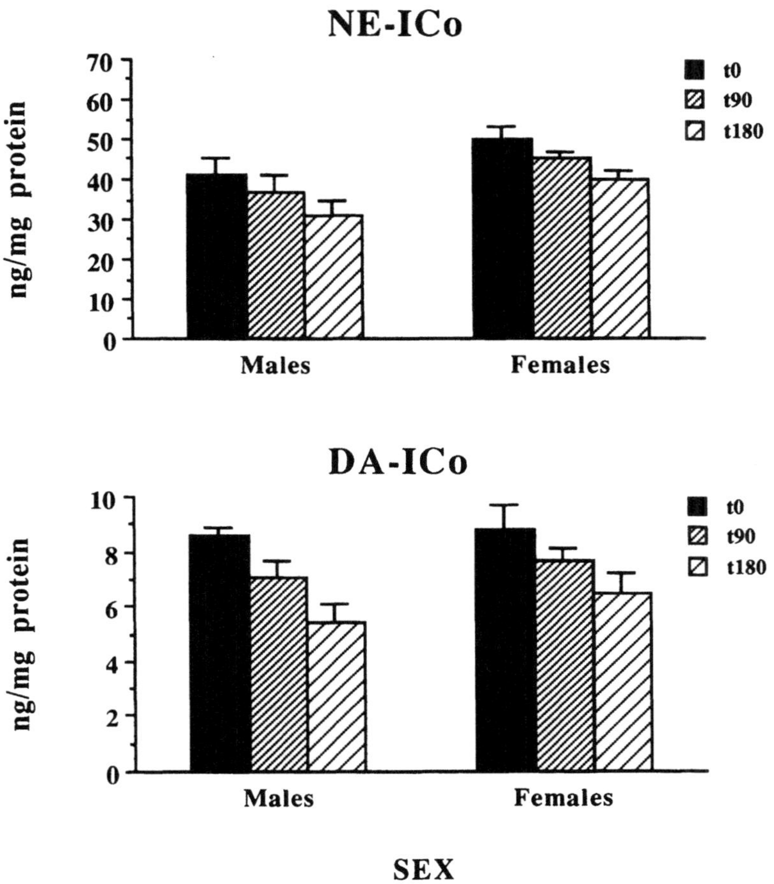

Fig. 7. *Continued*

regulated by gonadal hormones (J. Balthazart and G.F. Ball, unpubl. data).

In the ICo, there is a high level of binding apparent for both PAC and NMS (Ball et al. 1989a,b, 1990; Ball and Balthazart 1990), and there was a moderate level of PRZ binding that highlighted the nucleus relative to adjacent areas such as the nucleus mesencephalicus lateralis pars dorsalis (MLd) (Balthazart et al. 1989a). The density of binding sites labeled by both [^{3}H] PAC and [^{3}H] NMS changed significantly as one progressed in the rostral-caudal plane. Both the qualitative and quantitative pattern of binding differed between the two ligands (Ball and Balthazart 1990). Therefore, it was necessary to assess sex differences in the receptor density and possible steroid effects on receptor binding separately in different parts of the nucleus. In Fig. 8, the qualitative pattern of receptor density for the two ligands is presented schematically in the transverse plane.

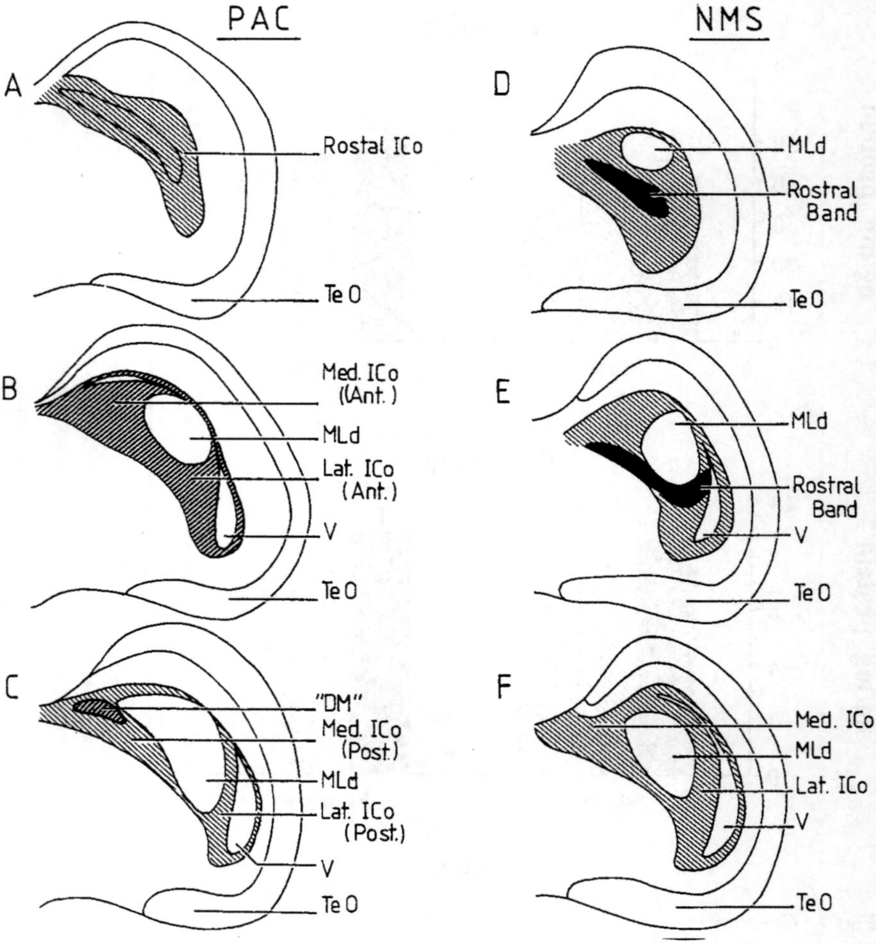

Fig. 8A–F. Qualitative pattern of [^{3}H] *para*-aminoclonidine (*PAC*) and [^{3}H] *n*-methylscopolamine (*NMS*) binding in the nucleus intercollicularis (*ICo*) of the Japanese quail. Sections **A** through **C** and **D** through **F** are arranged in a rostral to caudal order. For a given ligand, the darkest regions correspond to the areas displaying the heaviest binding. *Ant* Anterior; *DM* dorsomedial part of the ICo; *Lat* lateral; *Med* medial; *MLd* nucleus mesencephalicus lateralis pars dorsalis; *Post* posterior; *TeO* tectum opticum; *V* ventricle. (After Ball and Balthazart 1990)

The highest density of cholinergic receptors was detected in the most rostral extent of the ICo where the nucleus consists of a thick elongated band of cells whose long axis runs in the medial to lateral plane (referred to in the figures as the "rostral band"). The MLd is not yet apparent at the most rostral extent of this band, or if it is, it is only just visible and is smaller than it will appear when one is sampling at a more caudal level, as illustrated in Fig. 8D and E; the dense band of NMS binding always appears ventral to the MLd. As one samples caudally, the dense band of binding disappears, and the ICo appears as a teardrop consisting of two dense caps that surround the ovoid MLd (Fig. 8F). The NMS receptor density in both the lateral and the medial portions of the nucleus at this level is lower than that detected in the more anterior band of high binding, but it is higher than that measured in the adjacent MLd.

The density of α_2-adrenergic receptors also changes as one samples throughout the rostral to caudal extent of the ICo. At the most rostral extent of the nucleus (Fig. 8A) where it consists of an elongated band of cells without the MLd being apparent or with the MLd at its most rostral extent and therefore very small, the density of [^{3}H] PAC binding is consistent throughout the nucleus in contrast to what was described above for [^{3}H] NMS binding, which heavily labels an elongated ventral subregion of the nucleus. Caudal to this level, when the nucleus surrounds the entire MLd, the density of [^{3}H] PAC binding sites for both the medial and lateral portions of the nucleus is significantly higher than that observed at the more rostral level (Fig. 8B). This density declines as one samples more caudally in both parts of the nucleus. The one exception to this decline in receptor density is the dorsomedial subregion of the nucleus that is apparent at this level (labeled "DM" in Fig. 8C). This subregion can also be differentiated from the other parts of the ICo by the higher density of cells containing immunoreactive estrogen receptors (Ball et al. 1989b).

The density of [^{3}H] PAC binding in the MLd at both the rostral and caudal levels of the nucleus illustrated in Fig. 8 is low relative to the density in the surrounding ICo. It should also be noted that at the most caudal level illustrated in Fig. 8B and C, the ICo as defined by [^{3}H] PAC binding differs from the pattern revealed using [^{3}H] NMS. The relatively low binding area in the center of the ICo-MLd complex that generally corresponds to the MLd is much larger at this level when one compares a [^{3}H] PAC autoradiogram with a [^{3}H] NMS autoradiogram. It thus appears that only the outer edges of the ICo are being labelled by [^{3}H] PAC at this very caudal portion of the nucleus. Selected autoradiograms are presented in Fig. 9 to illustrate this differential binding of [^{3}H] NMS and [^{3}H] PAC in the ICo-MLd complex.

No sex difference in the [^{3}H] NMS distribution was detected in any portion of the ICo, and a significant effect of treatment was only detected in the rostral band of heavy binding (Fig. 10). This of course suggests that additional studies on sex differences and organizational effects of steroids might be useful to understand better the regulation of muscarinic receptors in this nucleus. When [^{3}H] PAC binding was considered in the entire ICo, the binding in males was found to be significantly higher than that apparent in females (Ball et al. 1989a). When this binding was subdivided by region, the difference was found to be attributable primarily to the lateral part of the nucleus (Ball and Balthazart 1990). A significant effect of steroid treat-

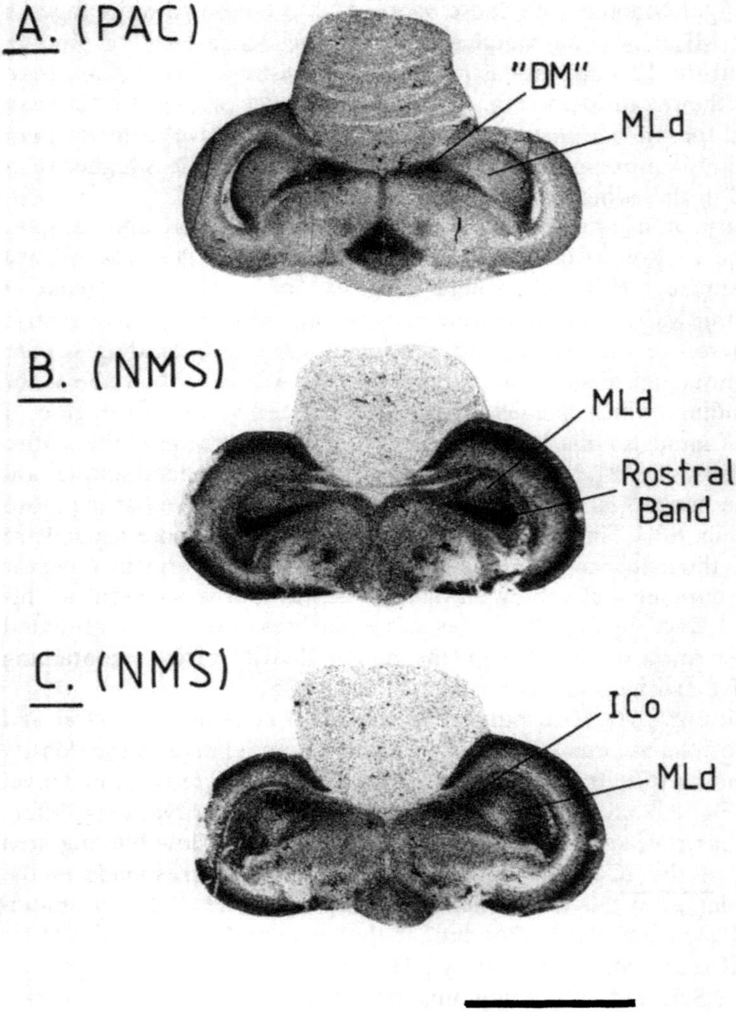

Fig. 9A–C. Autoradiograms illustrating the [^{3}H] *para*-aminoclonidine (*PAC; A*) and [^{3}H] *n*-methylscopolamine (*NMS*; **B C**) binding in the nucleus intercollicularis (*ICo*) in male quail. *A* approximately corresponds to the level labeled C in Fig. 8, and **B** and **C** correspond to levels D and F in Fig. 8. "*DM*" dorsomedial part of ICo; *MLd* nucleus mesencephalicus lateralis pars dorsalis. *Magnification bar* in 5 mm. (Based on data from Ball et al. 1989a,b, 1990)

ment was observed in the medial portion of the ICo at the level identified in Fig. 8B, and there was a significant effect of treatment and a significant interaction of treatment and sex in the lateral portion of the ICo at the same level (see Fig. 10). The receptor density was quantified in the MLd and all other subregions of the nucleus as defined based on the pattern of binding

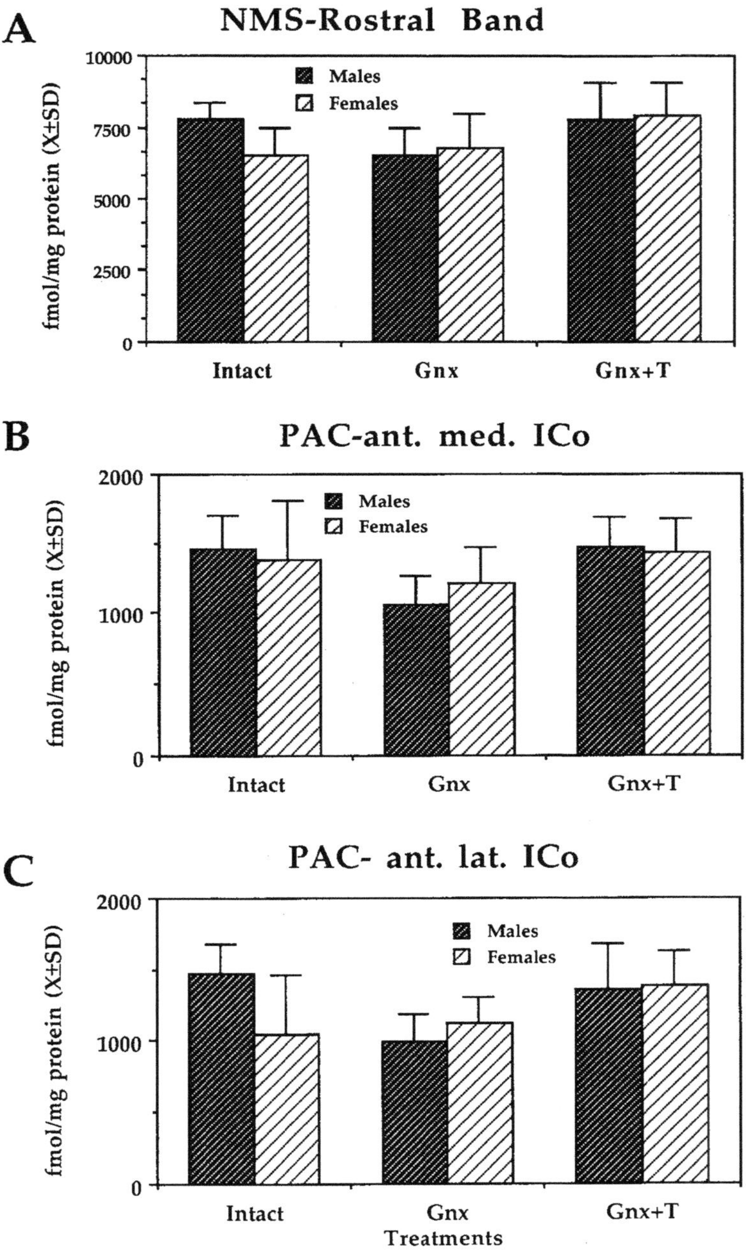

Fig. 10A–C. Effect of gonadectomy and testosterone treatment on [^{3}H] *n*-methylscopolamine (*NMS*) and [^{3}H] *para*-aminoclonidine (*PAC*) binding in subregions of the nucleus intercollicularis (*ICo*) in male and female quail. Birds of both sexes were studied as sexually mature animals (*Intact*) or after gonadectomy (*Gnx*) followed in some animals by a replacement therapy with testosterone (*Gnx + T*). See Fig. 8 for the definition of the ICo subregions. *Ant* Anterior; *Lat* lateral; *Med* medial. (After Ball and Balthazart 1990)

for both ligands. In no case was any effect of gonadectomy or steroid treatment on the binding density of either [^{3}H] NMS or [^{3}H] PAC detected in these other regions.

4 Discussion

Successful sexual reproduction requires the coordination of a variety of physiological events such as gonadal recrudescence and the growth of secondary sexual characteristics (Wingfield and Farner 1980), as well as the behavioral coordination between the two mating animals (Hinde 1965; Lehrman 1965). In the Japanese quail, as is true for many other seasonally breeding birds, reproductive activity is first triggered as a result of increasing daylength in the spring (Follett 1984). Photostimulation upregulates the secretion by the anterior pituitary of the two gonadotropin hormones, luteinizing hormone (LH) and follicle-stimulating hormone (FSH) (Follett 1976; Balthazart et al. 1979). These two hormones in turn regulate the secretion of sex steroids, such as T and E_2, by the gonads. In males, T is a potent signal for the initiation and coordination of a wide variety of reproductive events. A number of reproductive behaviors are dependent on an increased secretion of T. In some avian species, the various behaviors associated with reproduction are spread over several months during the spring and summer breeding period. For example, in many migratory species, the male will arrive well in advance of the female and establish a breeding territory and defend it from other conspecifics (Wingfield 1985). Several weeks or months later he will actively court a female, build a nest, and copulate with his mate. All these aspects of the behavioral repertoire, in most species, are activated to some extent by T (Silver et al. 1979). The physiological factors mediating these behavioral transitions from one breeding behavior to another, such as territory defense to courtship and/or nest building are not well understood at present. However, it has been argued that in male ring doves (*Streptopelia risoria*), the transition from aggressive courtship behavior to nest-oriented courtship is mediated by a T-induced increase in the preoptic aromatase activity (Hutchison and Steimer 1983). The higher levels of estrogen formed locally preferentially activate the nest-oriented courtship, which progressively replaces the aggressive courtship behavior which was triggered by T itself.

In the Japanese quail, crowing, the androgen-dependent vocalization used by the male to attract and stimulate females, and other reproductive behaviors such as copulation occur in a relatively close temporal sequence. For example, in laboratory housed quail, crowing, mounting, and copulation will be observed within a short (e.g., 10 min) observation period. It is therefore unlikely that neurochemical changes resulting from the action of a steroid hormone at the genomic level mediate these moment to moment alternations in behavior. The proximate factor immediately regulating the occurrence of these behaviors is the stimulus situation in which the male is placed. Males will predominantly crow when in visual isolation from conspecifics. When a female is introduced, they will switch to engaging in courtship behaviors such as strutting, which culminates in the actual copulation sequence, i.e., neck-grab, mounting, and cloacal apposition. The neurochemical changes induced by T in the POM and ICo, which are apparently associated with the activation of copulation and crowing, respectively, should not be considered to be the factors responsible for the alternation between one behavior and another. Rather, these changes represent the neurochemical environment that is an essential prelude for the occurrence of these behaviors in response to the appropriate stimuli. Long day lengths induce an increase in T secretion well in advance of the reproductive period. Among the many actions of T at this time is the induction of the appropriate neurochemical environment for the exhibition of reproductive behavior.

It is well-known that T has a diversity of effects depending on the target tissue considered. This is perhaps best illustrated in the most heterogeneous tissue in an organism, its brain. The present review has focused on two target areas that are thought to mediate two different behaviors necessary for successful reproduction. Important differences in the fate of T and the neurochemical consequences of its action in the POM and ICo have been identified. In the POM, T appears to act predominantly through its conversion to estrogenic metabolites such as estrone and/or E_2 (aromatization). The activity of the converting enzyme, aromatase, is up-regulated by T and also by its product E_2, at least in birds. The estrogens that are produced locally in the POM presumably bind with the estrogen receptors that are known to be present in the vast majority of neurons in the POM (Balthazart et al. 1989b; Watson and Adkins-Regan 1989a). In the ICo, aromatase is undetectable using both radioenzymatic and immunocytochemical methods. However, both estrogen and androgen receptors are common throughout this nucleus. Therefore, in contrast to the POM, T action in the ICo is the result of the activation of androgen receptors by T itself or by the androgenic metabolite 5α-DHT. The estrogen receptor appears to be stimulated directly by estrogens secreted in the periphery. Major neurochemical differences between the POM and ICo have also been described. The concentrations of both norepinephrine and dopamine are significantly higher in the POM than in the ICo, although the apparent turnover of norepinephrine (measured by the decrease in concentration following the administration of αMPT) might be higher in the latter. Sex differences in the baseline concentration and in the turnover of dopamine are seen in the POM but not in the ICo.

When considering the receptors for these and other transmitters, additional contrasts can be made between the two nuclei. The muscarinic cholinergic receptor is present in high concentrations in specific subregions of the ICo; it is nearly undetectable throughout the POM. α_1-Adrenergic receptors are also undetectable in the POM, while they are present at moderate densities relative to other areas of the quail brain in the ICo. Ligands binding to the α_2-adrenergic receptor label both nuclei. The pattern of receptor density is much more heterogeneous in the ICo than in the POM where there is a high level of receptor density apparent throughout the nucleus. T appears to regulate the density of the [^{3}H]NMS and [^{3}H]PAC binding sites in the ICo but not in the POM. The regulation of PRZ binding by steroids has not been investigated.

It is not known at present what specific aspects of the neurochemical environment in these two brain nuclei are related to the exhibition of the reproductive behaviors under consideration here. The noradrenergic system appears to inhibit male copulatory behavior since the selective noradrenergic neurotoxin DSP-4 has been shown enhance T-induced male copulatory behavior (Balthazart et al. 1988). In this respect, the higher concentration of norepinephrine in the POM in the female as compared with the male may explain why T is unable to activate copulatory responses in females. Similarly, the high baseline concentrations of dopamine and the high dopamine turnover in the male POM as compared with the female POM may play a causal role in the behavioral sex differences, because it is known, at least in mammals, that dopamine stimulates male sexual behavior (Meyerson et al. 1975; Bitran et al. 1988). The induction of aromatase by T appears to be a prerequisite and limiting step for the activation of copulatory behavior (Balthazart et al. 1990e). Although the aromatase activity also appears to be sexually dimorphic in the POA, this is not sufficient to explain the behavioral sex difference because the treatment of females with estrogen (a step which will bypass the role of the rate-limiting enzyme) does not effectively stimulate copulatory behavior (Schumacher and Balthazart 1983).

In the ICo, there is a high activity of 5α-reductase, which is consistent with the fact that crowing is activated by androgens and not estrogens. Noradrenaline acting through the α-adrenergic receptors is known to modulate vocalizations in birds and mammals (Barclay and Cheng, 1985; Kehoe 1988; Barclay 1989). This suggests that the presence of T-regulated α_2-receptors in the ICo may be involved in the control of vocal behaviors, such as crowing, in the quail. The existence of a T-regulated muscarinic receptor cannot be related in a precise way to the behavioral control of vocalizations at present.

The comparative study of hormone-neurochemical interactions in different nuclei of the brain of a single individual will permit us to evaluate the generality of conclusions that can be drawn from studies of a single model system such as the ventromedial nucleus of the hypothalamus as the site of the modulation of feminine sexual behavior by estrogen. Studying a single model system leaves unanswered a number of questions such as: Does the same steroid always produce the same neurochemical effect, or does it have different neurochemical consequences based on its site of action or the behavioral system it is regulating? Do the different effective steroid metabolites always activate the same biochemical process, or are they capable of activating different ones in different parts of the brain? At what point in the chain of biochemical steps leading to the activation of behavior does the identity of these different systems matter for the biological response? Is it at the level of the steroid circulating in the plasma, of the metabolite, of the steroid receptor to which the endogenous ligand binds, of the neurotransmitter or neuromodulator that is regulated by the hormone, or finally at the genomic level? To what degree are nongenomic effects involved in the mediation of the steroid effects on behavior? Do such nongenomic effects act in synergy with the genomic ones, or are they independent?

We are in a position to answer only a subset of these questions at present. Circulating steroids in the plasma provide a general signal for an increase in reproductive activity that is then modified and amplified at the various targets. In the brain, it seems clear that the type of metabolite produced provides specificity for the type of behavior response required. T metabolized to 5α-DHT induces crowing, but T aromatized to E_2 activates copulation. Based on the quail data provided here, it is impossible to evaluate whether the same metabolites of T will necessarily produce the same neurochemical changes in the brain. This has already been shown not to be the case in other animal species. For example, E_2 increases the progesterone receptor concentration in some brain areas, but not in others. Muscarinic cholinergic, β-adrenergic, and serotonin type 1 receptors are up-regulated by E_2 priming in ovariectomized female rats (McEwen et al. 1984). However, detailed regional analyses revealed that only a subclass of estrogen-sensitive sites will show increases in receptor density for these different systems. The pattern is unique for each of the different systems of receptors. It is thus clear that the same steroid can have different effects depending on the brain area considered. It is, however, at present impossible to discern whether any generalizations concerning the effects of different steroid metabolites on these various receptor subtypes will be possible. Furthermore, there is no logical reason to expect that there might be any limitation or rule as to how and which steroids can modify neurochemical functioning.

There is no particular level in the causal chain initiated by steroid action in the brain which determines in and of itself what kind of behavior will be activated. It will always be possible to find multiple behaviors influenced by the same metabolite and different behavioral systems that share at least a subset of T-modulated neurochemical events. In other words, there is no single neurochemical process that is specific to a given behavior. It is only the combination of the neural circuit and the specific trasmitters and modulators contained in the circuit which determines the behavioral "program" initiated by steroid action.

It had been assumed for many years that most if not all effects of steroids in the brain were mediated by the interaction with nuclear receptors, similar to what has been described in peripheral tissues such as the placenta, ovary, or seminal vesicles. Most effects of steroids on neurochemistry are thought to involve such genomic actions, and their rate of induction is consistent with this hypothesis. In addition to these receptor-mediated actions of steroids, recent work has suggested that membrane actions of steroids can affect such properties as neurotransmission and membrane composition (Naftolin et al. 1990b; Schumacher 1990). In the quail, we have recently identified immunoreactive aromatase on synaptic vesicles in the presynaptic boutons (Naftolin et al. 1990a). This raises the intriguing possibility that estradiol produced at the level of a synapse might act non-genomically at the membrane level. This idea would be consistent with the finding that most of the aromatase-immunoreactive cells in the POM do not contain immunoreactive estrogen receptors. The exact role played by E_2 at the membrane level is not directly related to the induction of copulation by T since this effect takes several days to a week to occur. It is, however, possible that a short-term modulation of behavior is mediated by these membrane effects.

The distribution of sex steroid receptors appears to be quite a conservative trait and suggests a remarkable homology among vertebrates (Morrell et al. 1975; Pfaff 1976; Kelley and Pfaff 1978; Morrell and Pfaff 1978). Steroid-metabolizing enzymes are also present in all major classes of vertebrates. However, the distribution of the metabolizing enzymes seems to have changed over the course of evolution in the vertebrate line (Callard et al. 1978a,b; Callard 1984). More primitive forms such as the teleosts show a very widespread distribution of aromatase activity in the brain. In this taxa, aromatase is present in relatively high concentrations in many parts of the brain that contain no estrogen receptors. As one compares the amphibians, reptiles, birds, and mammals, the number of areas in the brain that contain aromatase become fewer and fewer. In amniote vertebrates, the metabolizing enzymes are limited to areas of the limbic system, all of which are steroid-sensitive. It thus seems that the specialization of the vertebrate brain has involved an increased selectivity in the neuroanatomical sites at which T is aromatized into E_2. Aromatization seems to be limited to areas in which E_2 receptors are present and the locally produced estrogens can potentially act genomically. It should be noted, however, that within some of these areas, the estrogen receptors and aromatase are not generally colocalized (see above). In the quail, it seems possible that, at one time in phylogeny, the areas homologous to the ICo displayed the ability to aromatize T. As a part of the specialization to mediate different activities during reproduction, they have lost this ability. It seems plausible to suggest that a number of other neurochemical factors over time have changed in specific brain nuclei to facilitate differential activation by T. The occurrence of steroid receptors is in general evolutionarily conservative, except in certain taxa with unique specializations such as the songbirds (Ball 1990), but other components of the causal chain underlying steroid action, including the structure of the particular steroid circulating in the plasma in some cases, have changed. One of the challenges that need to be faced for the development of a complete theory of steroid action in the brain will be to ascertain whether or not there are any constraints on what brain processes can be modulated by steroids and if there are any rules that will explicate the bewildering diversity of steroid-neurochemical interactions that have been described.

Acknowledgements. The work reported in this paper was supported by grants from the National Institutes of Health, Bethesda, MD (HD 22064), the Belgian Fonds de la Recherche Fondamental Collective (9.4601.90 and 2.9003.91), the University of Liège (Fonds Spéciaux pour la Recherche 1990) and the EEC (SC1-0230-C/TT) to J. Balthazart. The collaboration between J. Balthazart and G.F. Ball was supported by a NATO grant (CRG910526). We would like to thank here all our recent collaborators and especially Michael Schumacher, Agnès Foidart, and Chantal Surlemont who have been of invaluable help in collecting all the data summarized in this review. We would like to thank Dr. M.M. McCarthy for useful discussions.

References

Adkins EK (1975) Hormonal basis of sexual differentiation in the Japanese quail. J Comp Physiol Psychol 89:61–71

Adkins EK (1977) Effects of diverse androgens on the sexual behavior and morphology of castrated male quail. Horm Behav 8:201–207

Adkins EK, Adler NT (1972) Hormonal control of behavior in the Japanese quail. J Comp Physiol Psychol 81:27–36

Adkins EK, Pniewski EE (1978) Control of reproductive behavior by sex steroids in male quail. J Comp Physiol Psychol 92:1169–1178

Adkins EK, Boop JJ, Koutnik DL, Morris JB, Pniewski EE (1980) Further evidence that androgen aromatization is essential for the activation of copulation in male quail. Physiol Behav 24:441–446

Adkins-Regan E (1983) Sex steroids and the differentiation and activation of avian reproductive behaviour. In: Balthazart J, Gilles R (eds) Hormones and behaviour in higher vertebrates. Springer, Berlin Heidelberg New York, pp 219–228

Alexandre C, Balthazart J (1986) Effects of metabolism inhibitors, antiestrogens and antiandrogens on the androgen and estrogen induced sexual behavior in Japanese quail. Physiol Behav 38:581–591

Armitage SE, Seller TJ (1981) Midbrain regions involved in call production of Japanese quail. Experientia 37:847–848

Ball GF (1990) Chemical neuroanatomical studies of the steroid-sensitive songbird vocal control system: a comparative approach. In: Balthazart J (ed) Hormones, brain and behaviour in vertebrates. 1. Sexual differentiation, neuroanatomical aspects, neurotransmitters and neuropeptides. (Comp Physiol, vol 8) Karger, Basel, pp 148–167

Ball GF, Balthazart J (1990) Steroid modulation of muscarinic cholinergic and α_2-adrenergic receptor density in the nucleus intercollicularis of the Japanese quail. Eur J Neurosci 2:828–835

Ball GF, Nock B, McEwen BS, Balthazart J (1989a) Distribution of α_2-adrenergic receptors in the brain of the Japanese quail as determined by quantitative autoradiography: implications for the control of sexually dimorphic reproductive processes. Brain Res 491:68–79

Ball GF, Foidart A, Balthazart J (1989b) A dorsomedial subdivision within the nucleus intercollicularis identified in the Japanese quail (*Coturnix coturnix japonica*) by means of α_2-adrenergic receptor autoradiography and estrogen receptor immunohistochemistry. Cell Tissue Res 257:123–128

Ball GF, Nock B, Wingfield JC, McEwen BS, Balthazart J (1990) Muscarinic cholinergic receptors in the songbird and quail brain: a quantitative autoradiographic study. J Comp Neurol 298:431–442

Balthazart J (1989) Steroid metabolism and the activation of social behavior. In: Balthazart J (ed) Advances in comparative and environmental physiology, vol 3. Springer, Berlin Heidelberg New York, pp 105–159

Balthazart J, Ball GF (1989) Effects of the noradrenergic neurotoxin DSP-4 on luteinizing hormone levels, catecholamine concentrations, α_2-adrenergic receptor binding, and aromatase activity in the brain of the Japanese quail. Brain Res 492:163–175

Balthazart J, Surlemont C (1990a) Copulatory behavior is controlled by the sexually dimorphic nucleus of the quail POA. Brain Res Bull 25:7–14

Balthazart J, Surlemont C (1990b) Androgen and estrogen action in the preoptic area and activation of copulatory behavior in quail. Physiol Behav 48:599–609

Balthazart J, Massa R, Negri-Cesi P (1979) Photoperiodic control of testosterone metabolism, plasma gonadotrophins, cloacal gland growth, and reproductive behavior in the Japanese quail. Gen Comp Endocrinol 39:222–235

Balthazart J, Schumacher M, Ottinger MA (1983) Sexual differences in the Japanese quail: behavior, morphology and intracellular metabolism of testosterone. Gen Comp Endocrinol 51:191–207

Balthazart J, Schumacher M, Malacarne G (1985) Interaction of androgens and estrogens in the control of sexual behavior in male Japanese quail. Physiol Behav 35:157–166

Balthazart J, Delville Y, Sulon Y, Hendrick JC (1987) Plasma levels of luteinizing hormone and of five steroids in photostimulated, castrated and testosterone-treated male and female Japanese quail (*Coturnix coturnix japonica*). Gen Endocrinol (Life Sci Adv) 5:31–36

Balthazart J, Libioulle JM, Sante P (1988) Stimulatory effects of the noradrenergic neurotoxin DSP4 on sexual behavior in male quail. Behav Processes 17:27–44

Balthazart J, Ball GF, McEwen BS (1989a) An autoradiographic study of α_1-adrenergic receptors in the brain of the Japanese quail (*Coturnix coturnix japonica*). Cell Tissue Res 258:563–568

Balthazart J, Gahr M, Surlemont C (1989b) Distribution of estrogen receptors in the brain of the Japanese quail: an immunocytochemical study. Brain Res 501:205–214

Balthazart J, Evrard L, Surlemont C (1990a) Effects of the non-steroidal aromatase inhibibor, R76713 on testosterone-induced sexual behavior in the Japanese quail (*Coturnix coturnix japonica*). Horm Behav 24:510–531

Balthazart J, Foidart A, Harada N (1990b) Immunocytochemical localization of aromatase in the brain. Brain Res 514:327–333

Balthazart J, Foidart A, Surlemont C, Harada N, Leranth C, Naftolin F (1990c) Immunocytochemical localization of aromatase and estrogen receptors in the brain. Soc Neurosci 16:1313 (Abstr 542–13)

Balthazart J, Foidart A, Surlemont C, Harada N (1990d) Aromatase is generally not colocalized with estrogen receptors in the brain. J Neurobiol 22:143–157

Balthazart J, Foidart A, Hendrick JC (1990e) The induction by testosterone of aromatase activity in the preoptic area and activation of copulatory behavior. Physiol Behav 47:83–94

Balthazart J, Schumacher M, Evrard L (1990f) Sex differences and steroid control of testosterone-metabolizing enzymes activity in the quail brain. J Neuroendocrinol 2: 675–683

Balthazart J, Sante P, Ball GF (1991) Testosterone effects on the staining density and autoradiographic investigations of the α_2-adrenergic receptor in the medial preoptic nucleus of the Japanese quail: relationship to the activation of reproductive behavior. Arch Int Physiol Biochim 99:385–392

Barclay SR (1989) The role of the noradrenergic system in male courtship behavior of the ring dove. Doctoral dissertation, Rutgers University, Newark, NJ

Barclay SR, Cheng MF (1985) The role of alpha adrenergic system in the male ring dove's courtship behavior. Soc Neurosci 11:736 (Abstr)

Baylé JD, Ramade F, Oliver J (1974) Stereotaxic topography of the brain of the quail. J Physiol (Paris) 68:219–241

Beach FA (1948) Hormones and behavior. Hoeber, New York

Bitran D, Hull EM, Holmes GM, Lookingland KJ (1988) Regulation of male copulatory behavior by preoptic incertohypothalamic dopamine neurons. Brain Res Bull 20:323–331

Blaustein JD, Olster DH (1989) Gonadal steroid hormone receptors and social behaviors. In: Balthazart J (ed) Advances in comparative and environmental physiology, vol 3. Springer, Berlin Heidelberg New York, pp 31–104

Callard GV (1984) Aromatization in brain and pituitary: an evolutionary perspective. In: Celotti F, Naftolin F, Martini L (eds) Metabolism of hormonal steroids in the neuroendocrine structures. Raven Press, New York, pp 79–102

Callard GV, Petro Z, Ryan KJ (1978a) Phylogenetic distribution of aromatase and other androgen-converting enzymes in the central nervous system. Endocrinology 103:2283–2290

Callard GV, Petro Z, Ryan KJ (1978b) Conversion of androgen to estrogen and other steroids in the vertebrate brain. Am Zool 18:511–523

Carter CS, Getz LL (1985) Social and hormonal determinants of reproductive patterns in the prairie vole. In: Gilles R, Balthazart J (eds) Neurobiology. Springer, Berlin Heidelberg New York, pp 18–36

Cohen J (1981) Hormones and midbrain mediation of courtship behavior in the male ring dove. J Comp Physiol Psychol 95:512–528

Cohen J (1983) Hormones and brain mechanisms of vocal behaviour in non-vocal learning birds. In: Balthazart J, Pröve E, Gilles R (eds) Hormones and behaviour in higher vertebrates. Springer, Berlin Heidelberg New York, pp 422–436

Cohen J, Cheng MF (1981) The role of the midbrain in courtship behavior of the female ring dove (*Streptopelia risoria*): evidence from radiofrequency lesion and hormone implant studies. Brain Res 207:279–301

Cohen J, Cheng MF (1982) Effects of testosterone metabolites and estrogen in the midbrain control of courtship behavior in the male ring dove (*Streptopelia risoria*). Neuroendocrinology 34:64–74

Dohanich G, Nock B, McEwen BS (1985) Steroid hormones, receptors and neurotransmitters. In: Moudgil VK (ed) Molecular mechanism of steroid hormone action. deGuyter, Berlin, pp 701–732

Feder HH (1984) Hormones and sexual behavior. Annu Rev Psychol 35:165–200

Foidart A, Surlemont C, Ball GF, Balthazart J (1990) Neurochemical differences in steroid-sensitive areas mediating reproductive behaviors in quail. Soc Neurosci 16:753 (Abstr 311–14)

Follett BK (1976) Plasma follicle-stimulating hormone during photoperiodically induced sexual maturation in male Japanese quail. J Endocrinol 69:117–126

Follett BK (1984) Birds. In: Lamming GE (ed) Marshall's physiology of reproduction. Longman Green, Edinburgh, pp 283–350

Gardner A, Harding CF (1990) Steroid modulation of choline acetyltransferase activity in the zebra finch brain. Soc Neurosci 16:922 (Abstr 380–5)

Guyomarc'h JC (1974) Influence du benzoate d'œstradiol sur la fréquence et la hauteur des cris de contact chez la caille Japonaise. Bull Soc Zool Fr 4:743–751

Harlan RE (1988) Regulation of neuropeptide gene expression by steroid hormones. Mol Neurobiol 2:183–200

Hinde RA (1965) Interaction of internal and external factors in integration of canary

reproduction. In: Beach FA (ed) Sex and behavior. Wiley, New York, pp 381–415
Hutchison JB, Steimer T (1983) Hormone-mediated behavioural transitions: a role for brain aromatase. In: Balthazart J, Prove E, Gilles R (eds) Hormones and behaviour in higher vertebrates. Springer, Berlin Heidelberg New York, pp 261–274
Johnson AE, Nock B, McEwen BS, Feder HH (1985) Estradiol modulation of alpha 2-noradrenergic receptors in guinea pig brain assessed by tritium-sensitive film autoradiography. Brain Res 336:153–157
Johnson AE, Nock B, McEwen BS, Feder HH (1988) α_1- and α_2-Noradrenergic receptors in steroid-sensitive brain areas: development and response to estradiol-17β benzoate in neonatal guinea pigs. Dev Brain Res 42:247–252
Kehoe P (1988) Ontogeny of adrenergic and opioid effects on separation vocalizations in rats. In: Newman JD (ed) The physiological control of mammalian vocalization. Plenum Press, New York, pp 301–320
Kelley DB, Pfaff DW (1978) Generalizations from comparative studies on neuroanatomical and endocrine mechanisms of sexual behaviour. In: Hutchison JB (ed) Biological determinants of sexual behaviour. Wiley, Chichester, pp 225–254
Lehrman DS (1965) Interaction between internal and external environments in the regulation of the reproductive cycle of the ring dove. In: Beach FA (ed) Sex and behavior. Wiley, New York, pp 355–380
Luttge WG (1979) Endocrine control of mammalian male sexual behavior: an analysis of the potential role of testosterone metabolites. In: Beyer C (ed) Endocrine control of sexual behavior. Raven Press, New York, pp 341–363
Luttge WG, Hall NR (1973) Androgen-induced agonistic behavior in castrate male Swiss-Webster mice: comparison of four naturally occurring androgens. Behav Biol 8:725–732
McEwen BS, Pfaff DW (1985) Hormone effects on hypothalamic neurons: analysing gene expression and neuromodulator action. TINS 8:105–110
McEwen BS, Biegon A, Fischette CT, Luine VN, Parsons B, Rainbow TC (1984) Toward a neurochemical basis of steroid action. In: Martini L, Ganong WF (eds) Frontiers in neuroendocrinology. Raven Press, New York, pp 153–176
McEwen BS, Jones KJ, Pfaff DW (1987) Hormonal control of sexual behavior in the female rat: molecular, cellular and neurochemical studies. Biol Reprod 36:37–45
McEwen BS, Luine VN, Fischette CT (1988) Developmental actions of hormones: from receptors to function. In: Easter SS Jr, Barald KF, Carlson BM (eds) From message to mind. Sinauer Associates, Sunderland, MA, pp 272–287
Meyerson BJ, Palis A, Sietniks A (1975) Hormone-monoamine interactions and sexual behavior. In: Beyer C (ed) Endocrine control of behavior. Raven Press, New York, pp 389–404
Morrell JI, Pfaff DW (1978) A neuroendocrine approach to brain function: localization of sex steroid concentrating cells in vertebrate brains. Am Zool 18:447–460
Morrell JI, Kelley DB, Pfaff DW (1975) Sex steroid binding in the brain of vertebrates. In: Knigge KM, Scott DE, Kobayashi H, Miura S, Ishii S (eds) Brain-endocrine interactions II. Karger, Basel, pp 230–256
Naftolin F, Leranth C, Balthazart J (1990a) Ultrastructural localization of aromatase immunoreactivity in hypothalamic neurons. Endocr Soc (Abstr 669), p 192
Naftolin F, Garcia-Segura LM, Keefe D, Leranth C, MacLusky NJ, Brawer JR (1990b) Estrogen effects on the synaptology and neural membranes of the rat hypothalamic arcuate nucleus. Biol Reprod 42:21–28
Panzica GC, Viglietti-Panzica C, Calcagni M, Anselmetti GC, Schumacher M, Balthazart J (1987) Sexual differentiation and hormonal control of the sexually dimorphic preoptic medial nucleus in quail. Brain Res 416:59–68
Pfaff DW (1976) The neuroanatomy of sex hormone receptors in the vertebrate brain. In: Anand Kumar TC (ed) Neuroendocrine regulation of fertility. Karger, Basel, pp 30–45
Pfaff DW (1980) Estrogen and brain function. Springer, Berlin Heidelberg New York
Pfaff DW (1989a) Patterns of steroid hormone effects on electrical and molecular events in hypothalamic neurons. Mol Neurobiol 3:135–154

Pfaff DW (1989b) Features of a hormone-driven defined neural circuit for a mammalian behavior. Principles illustrated, neuroendocrine syllogisms, and multiplicative steroid effects. Ann NY Acad Sci 563:131–147

Pfaff DW, Schwartz-Giblin S (1988) Cellular mechanisms of female reproductive behaviors. In: Knobil E, Neill J (eds) The physiology of reproduction. Raven Press, New York, pp 1487–1568

Phillips RE, Barfield RJ (1977) Effects of testosterone implants in midbrain vocal areas of capons. Brain Res 122:378–381

Potash LM (1970) Vocalizations elicited by electrical brain stimulation in *Coturnix coturnix japonica*. Behaviour 31:149–167

Ryan S, Arnold AP (1981) Evidence for cholinergic participation in the control of bird song: acetylcholinesterase distribution and muscarinic receptor autoradiography in the zebra finch brain. J Comp Neurol 202:211–219

Schumacher M (1990) Rapid membrane effects of steroid hormones: an emerging concept in neuroendocrinology. TINS 13:359–362

Schumacher M, Balthazart J (1983) The effects of testosterone and its metabolites on sexual behavior and morphology in male and female Japanese quail. Physiol Behav 30:335–339

Schumacher M, Balthazart J (1985) Sexual differentiation is a biphasic process in mammals and birds. In: Gilles R, Balthazart J (eds) Neurobiology. Current comparative approaches. Springer, Berlin Heidelberg New York, pp 203–219

Schumacher M, Balthazart J (1986) Testosterone-induced brain aromatase is sexually dimorphic. Brain Res 370:285–293

Seller TJ (1981) Midbrain vocalization centers in birds. TINS 4:301–303

Silver R, O'Connell M, Saad R (1979) The effects of androgens on the behavior of birds. In: Beyer C (ed) Endocrine control of sex behavior. Raven Press, New York, pp 223–279

Simerly RB, Young BJ, Capozza MA, Swanson LW (1989) Estrogen differentially regulates neuropeptide gene expression in a sexually dimorphic olfactory pathway. Proc Natl Acad Sci USA 86:4766–4770

Vischer A, Cuénod M, Henke H (1982) Neurotransmitter receptor ligand binding and enzyme regional distribution in the pigeon visual system. J Neurochem 38:1372–1382

Watson JT, Adkins-Regan E (1980a) Neuroanatomical localization of sex steroid-concentrating cells in the Japanese quail (*Coturnix japonica*): autoradiography with [^{3}H]-testosterone, [^{3}H]-estradiol, and [^{3}H]-dihydrotestosterone. Neuroendocrinology 49:51–64

Watson JT, Adkins-Regan E (1989b) Activation of sexual behavior by implantation of testosterone propionate and estradiol benzoate into the preoptic area of the male Japanese quail (*Coturnix japonica*). Horm Behav 23:251–268

Watson JT, Abdelnabi M, Wersinger S, Ottinger MA, Adkins-Regan E (1990) Circulating estradiol and the activation of male-typical copulatory behavior in Japanese quail (*Coturnix japonica*). Gen Comp Endocrinol 77:229–238

Wingfield JC (1985) Environmental and endocrine control of territorial behavior in birds. In: Follett BK, Ishii S, Chandola A (eds) The endocrine system and the environment. Japan Sci Soc Press/Springer, Tokyo Berlin Heidelberg New York pp 265–277

Wingfield JC, Farner DS (1980) Control of seasonal reproduction in temperate-zone birds. Prog Reprod Biol 5:62–101

Subject Index

Printing: Mercedesdruck, Berlin
Binding: Buchbinderei Lüderitz & Bauer, Berlin

GPSR Compliance

The European Union's (EU) General Product Safety Regulation (GPSR) is a set of rules that requires consumer products to be safe and our obligations to ensure this.

If you have any concerns about our products, you can contact us on ProductSafety@springernature.com

In case Publisher is established outside the EU, the EU authorized representative is:

Springer Nature Customer Service Center GmbH
Europaplatz 3
69115 Heidelberg, Germany

Batch number: 10620141

Printed by Printforce, the Netherlands